Admitting the Holocaust

Admitting the HOLOCAUST

Collected Essays

LAWRENCE L. LANGER

New York
OXFORD UNIVERSITY PRESS
Oxford

Oxford University Press

Oxford New York
Athens Auckland Bangkok Bombay
Calcutta Cape Town Dar es Salaam Delhi
Florence Hong Kong Istanbul Karachi
Kuala Lumpur Madras Madrid Melbourne
Mexico City Nairobi Paris Singapore
Taipei Tokyo Toronto

and associated companies in
Berlin Ibadan

First published in 1995 by Oxford University Press, Inc.,
198 Madison Avenue, New York, New York 10016

First issued as an Oxford University Press paperback, 1996

Library of Congress Cataloging-in-Publication Data
Langer, Lawrence L.
Admitting the Holocaust : collected essays / Lawrence L. Langer.
p. cm. Includes bibliographical references.
ISBN 0-19-509357-7
ISBN 0-19-510648-2 (Pbk.)
1. Holocaust, Jewish (1939–1945)—Moral and ethical aspects.
2. Holocaust, Jewish (1939–1945)—Historiography.
3. Holocaust, Jewish (1939–1945), in literature.
4. Literature, Modern—20th century—History and criticism.
I. Title. D803.3L358 1994 940.53'18—dc20 94-13368

10 9 8 7 6 5 4 3 2 1

Printed in the United States of America

"The Americanization of the Holocaust on Stage and Screen," from *From Hester Street to Hollywood: The Jewish-American Stage and Screen,* Susan Blacher Cohen, ed. Copyright © 1983 by Indiana University Press. Reprinted by permission of the publisher.

"Beyond Theodicy: Jewish Victims and the Holocaust," reprinted from the journal, *Religious Education,* vol. 84, no. 1, by permission from the publisher, The Religious Education Association.

"Cultural Resistance to Genocide" by Lawrence Langer first appeared in *Witness,* vol. 1, no. 1, 1987, published by Oakland Community College, Bloomfield Hill, Michigan. Reprinted by permission of the publisher.

"Fictional Facts and Factual Fictions: History in Holocaust Literature," from *Reflections of the Holocaust in Art and Literature,* by Braham, Randolph L., © 1990 by Columbia University Press, New York. Reprinted with permission of the publisher.

"Kafka as Holocaust Prophet: A Dissenting View," from *Kafka's Contextuality,* Alan Udoff, ed. Copyright © 1986 by Alan Udoff.

"The Literature of Auschwitz" by Lawrence Langer from *The Anatomy of an Auschwitz Death Camp.* Copyright © 1994 by Yisrael Gutman and Michael Berenbaum. Reprinted by permission of Indiana University Press.

"Malamud's Jews and the Holocaust Experience" by Lawrence L. Langer. Reprinted with permission of G.K. Hall & Co., an imprint of Macmillan Publishing Company, from *Critical Essays on Bernard Malamud,* edited by Joel Salzberg. Copyright © 1987 by Joel Salzberg.

"Memory's Time: Chronology and Duration in Holocaust Testimonies," from *The Yale Journal of Criticism,* vol. 6, no. 2. © 1993 by Yale University.

"A Tainted Legacy: Remembering the Warsaw Ghetto." Reprinted from TIKKUN MAGAZINE, A BI-MONTHLY JEWISH CRITIQUE OF POLITICS, CULTURE, AND SOCIETY. 251 West 100th Street, 5th floor, New York, NY 10025.

"Understanding Atrocity: Killers and Victims in the Holocaust" first appeared in *Michigan Quarterly Review,* fall 1985, vol. 24, no. 1

For **Sandy**

who deserves them all

and

David and **Joshua**

who deserve one too

Preface

When I was in Munich more than a quarter of a century ago writing my first book, *The Holocaust and the Literary Imagination*, I asked various German friends what they thought was the most important recent work on the subject of the deathcamp experience. To my surprise, they all recommended the same title: Jean Améry's *Jenseits von Schuld und Sühne* (*Beyond Guilt and Atonement*, translated into English many years later as *At the Mind's Limits*). I found a copy the day before I left for America, and began reading it on the leisurely ship voyage home. Readers of the essays in this volume will see at once the vivid and lasting impression the views of this Auschwitz survivor have had on my response to the Holocaust experience. His line "No bridge led from death in Auschwitz to *Death in Venice*" must have been engraved on my memory as nothing less than an epiphany, so often does it recur in my own writing. Améry's discovery of the permanent rift between death in literature and mass murder by Nazi Germany is an insight whose implications still have not received full appreciation.

Shortly after my return—this was in 1969—I received a call from a young man who asked if he could talk to me about the Holocaust. He brought with him a copy of another book unfamiliar to me, Charlotte Delbo's *None of Us Will Return*, and I read this with the same eager sense of revelation with which I had consumed Améry's volume. Also a survivor of Auschwitz, Delbo had done for style what Améry had done for thought: noticed a breach in narrative traditions that demanded a totally new verbal sensibility to capture the past she had endured. Delbo's vision proved as indelible as Améry's ideas; her recurrent influence will also be evident to readers of these essays.

The third major source to have shaped my understanding of this event is the oral testimonies of Holocaust survivors, with which I have been involved for nearly a decade. The authority of their voices has kept me grounded in the daily stress of the time, and if I return to those voices often in these essays, it is because I am convinced that they offer us an access to the event as vital as the ones opened up for us by the achievements of the literary imagination. Looking back, I see now that my career has been divided among these inspirations. The Holocaust experience has made possible a literature of the Holocaust; such literature repays the debt by preventing the experience from receding to the distant frontiers of memory. The advance in my own work from *The Holocaust and the Literary Imagination* to *Holocaust Testimonies: The Ruins of Memory* represents less a progression than a dual commitment. The essays collected in *Admitting the Holocaust* reflect a shifting concern between the two—literature and testimony—during the past decade. I owe an immense debt to those who were determined to record for posterity the details of their often unbearable ordeals, and to the writers whose creative talents have helped us to imagine the unthinkable. Without their separate efforts, I would have had nothing to say.

West Newton, Mass.
May 1994

L.L.L

Contents

Admitting the Holocaust

Introduction

f the Holocaust has taught us anything, it is that we were other than we believed, masters of neither time nor space. Deportees were forced into zones of confinement void of a future, hostages only to death. Nazi Germany demoted its victims to the rank of less-than-human, turning them against their will into beings who found that choice was often an illusion. Armed with the weapons of hindsight and foresight, those *outside* the net still did little to thwart the catastrophe or help others to vault its barricades into a securer domain. They, too, lost control over time and space. Discouraged by these failures, we continue to resist their import by groping for ways to change them into limited versions of success. The result is a persisting myth about the triumph of the spirit that colors the disaster with a rosy tinge and helps us to manage the unimaginable without having to look at its naked and ugly face.

The essays in *Admitting the Holocaust* represent my attempt over nearly a decade to wrestle with the rupture in human values as it really was—a rupture that after the war left stunned minds staring blankly at alien modes of living and dying in the monstrous milieu of ghettos and camps. One response to this scenario might be to accept the Holocaust as a warrant for futility, a witness to the death of hope. But another, the one I choose to follow, is to regard the calamity as a summons to reconsider usual views of the self, its relation to time and memory, its portrayal in literature born of the Holocaust, its use and abuse by culture, and its role in reshaping our sense of history's legacies from the past and bequests to later generations. Such probing may not transform our present or even improve our prospects, and certainly will not redeem our past, but at least it can help us to live with clearer eyes. Clarity has a virtue of its own in murky times.

A writer like Bernard Malamud—whose response to the event I examine in one of these essays—found suffering a source of spiritual strength, much in the manner of Job. The Holocaust must have seemed a threat to his vision of man's basic dignity. His few literary forays into its domain were never entirely successful. Cynthia Ozick, on the contrary, remains one of the few American writers to meet the challenge of imagining mass murder without flinching; with great subtlety, as I try to explain, she distinguishes between Holocaust myth and Holocaust truth. For many critics unsympathetic to the idea of rupture and more comfortable with the custom of joining tradition to the individual talent, Kafka appeared as an obvious candidate for the title of precursor to the Holocaust—another issue I explore—as the need for nexus muted the risk of disruption. But finding links now between Kafka's world of menace and our own universe of atrocity raises the question of why Kafka's admirers—Walter Benjamin among them—did not foresee what lay before them then. The answer may be that Kafka has been graced with the title of prophet only retrospectively. Our blindness to the peril, compounded but not caused by indifference, seems to have launched an ex post facto effort to confirm connections not evident at the time. Perhaps this is a way of restoring pattern to an apparent chaos that flatters no one. Innate to culture is the search for correspondences: Could history, literature, religion, or morality do without it? Imaginations daunted by the unprecedented are eased by the reliance on precedents. Supplied with her Dostoevsky and her Russian dictionary, the Dutch Jew Etty Hillesum—another subject of these essays—regarded a trip to Westerbork or Auschwitz as a manageable challenge. How else could she have written in her diary, "[B]y excluding death from our life we cannot live a full life, and by admitting death into our life we enlarge and enrich it"? Inspired by one of her favorite authors, she allowed the vocabulary of martyrdom to shield her mind from the siege of liquidation.

And so do we, more than fifty years after the event. A serious student of the Holocaust experience recently described it as an ontological war against a metaphysical opponent. I shudder with dismay at such language, so obviously an evasion of the physical facts, because it substitutes a hypothetical trajectory for a particular target, like an archer praising the path of his arrow even though it falls wide of its mark. One aim of these essays is to enjoin the reader to focus on the bull's-eye and not to be distracted by the detours that avoid it. This requires a shift in thinking about ourselves and our cultural premises that I consider both necessary and dangerous—necessary, if we are to appreciate the exact nature of what we have done and suffered in our century, and dangerous because the adventure leads to a diminished sense of self that might further injure egos already damaged by the assaults of contemporary history. I suspect, however, that we have few other options, since we all have seen the fatal results of building a society on the fragile foundation of a naïve idealism that, when exposed, turns out to have been only a virtuous name for self-delusion.

Our age of atrocity clings to the stable relics of faded eras, as if ideas like natural innocence, innate dignity, the inviolable spirit, and the triumph of art over reality were immured in some kind of immortal shrine, immune to the ravages of history and time. Camus's caution that we learn to live within the limits of the possible seems a harsh sentence to a culture loath to surrender the promises of Romantic yearning. Victims and survivors of the Holocaust learned to redefine the meaning of possibility during their captive years, but after the ordeal was over, those who returned were inclined to adjust—who can blame them?—to "normal" pursuits and get on with their pillaged lives as best they could. As a result, the habit of discussing the past with a familiar discourse continues, while new models for dealing with mass murder intellectually, morally, historically, and philosophically do not proliferate.

The evidence for fresh vision is there, but the analysis is scanty. Commenting on the disbelief with which Aleksandr Solzhenitsyn's revelations in his Gulag Archipelago volumes were greeted—chiefly, that Stalin had left 60 million casualties in his wake—Russian poet Joseph Brodsky reflected: "I have a theory of why these things don't seep through, and that is a theory about self-preservation, mental self-preservation. Western man, by and large, is the most natural man, and he cherishes his mental comfort. It is almost impossible for him to admit disturbing evidence."[1] A world in which a relatively small number of men can cause the death of so many millions while screening their crimes and remaining themselves unpunished and unrepentant—for the most part, even after their defeat—is a world deprived of ethical force, one in which power supplants human concern, and indifference to suffering prevails over practical compassion.

The quest for mental comfort leads to a split that is as much verbal as it is emotional and intellectual. How else can we explain the subtitle that Yad Vashem in Jerusalem chose for its Holocaust memorial: Heroes and Martyrs Remembrance Authority? It established a tradition of avoidance that will be encouraged as long as the name exists, reminding us how much of our language about the Holocaust is designed to console instead of confront. Auschwitz is often described (with a certain melodramatic flair) as an "antiworld," a separate planet, though in fact the camp was in Poland in Europe on planet earth, run by men and women like the rest of us. They had more power and were prone to use it in ways that most of us would shrink from—or hope we would—but otherwise, with the exception of a handful of sadists, they did their work of destruction routinely and with detached satisfaction. Auschwitz wasn't the antiworld, but the world as Nazi Germany decided it should be. Removing its cosmology from our own by placing it in another universe is little more than an effort to sever ourselves from its acts and values in our ongoing search for what Brodsky calls mental comfort.

The role of language in this search—a subtext of many of these essays—illustrates how easy it is to change the impact of a disastrous event simply by renaming it. When we speak of the survivor instead of the victim and of martyrdom instead of murder, regard being gassed as a pattern for dying with dignity, or evoke the redemptive rather than the grievous power of memory, we draw on an arsenal of words that urges us to build verbal fences between the atrocities of the camps and ghettos and what we are mentally willing—or able—to face. When we greet "liberation" with relief and celebrate it as a victory for endurance, we block out the images of the emaciated bodies that Allied troops found as they entered places like Buchenwald and Bergen-Belsen, as if a single honorific term could erase the paradoxical possibility that survival might also be a lifelong sentence to the memory of loss. The concept of the "dead survivor" may seem contradictory, but for many of those who were still alive when the catastrophe ended, it is implicit in the narrative of their ordeal.

Kierkegaard wrote somewhere that life is lived forward and understood backward, and I suspect that for most of us this continues to be an attractive premise. But just the opposite is true for the Holocaust experience: we live it backward in time, and once we arrive there, we find ourselves mired in its atrocities, a kind of historical quicksand that hinders our bid to bring it forward again into a meaningful future. The American mind in particular—by heritage, training, and instinct—is unsympathetic to any forecast whose dominant image is an obstacle instead of a vista. Our culture teaches us to demolish barriers or clamber over them; to us, the death sentence is not a statement of human limits, but a feint of nature to be outfoxed by medical or religious tactics. Writer Harold Brodkey, when he learned that he was the victim of the fatal AIDS virus, commented painfully on how little support he received from his native milieu. An American daydream, he said, was about "rebuilding after the flood, about being better off than before, about outwitting this or that challenger, up to and including death."[2] Admitting the Holocaust into our national consciousness is the public equivalent of Brodkey's private dilemma.

The most undigestible fact of the Holocaust is that during its reign death outwitted us on such a massive scale that "being better off than before" ceased to be a viable consequence of the calamity. If we go on using a discourse of consolation about an event for which there is none, it is partly because old habits of language cling like burrs to the pelts of civilization, and partly because no full-fledged discourse of ruin, more appropriate to our hapless times, has yet emerged. A discourse of ruin may sound frightening, but only if we mistake it for a statement about the permanent defeat of hope. It simply asks us to be honest about the nature of the ruin, which means confessing that the Holocaust was not an illness from which its "patients" could be cured or a trauma from which victims "recovered." In his Nobel Prize acceptance speech many years ago, William Faulkner fervently declared, in a typically American sentiment, that man would not only endure, but also prevail. The discourse of ruin would drop the latter term in the sequence, since "prevailing over the Holocaust" is a manifesto without logic or meaning.

After the Holocaust, to say nothing of the other atrocities of our century, the future unfolds under a shadow of past disasters, and this requires a shift in how we imagine the burden and the promise of time, as well as many other concepts and images that once lent balance to our spiritual and psychological cosmos. Holocaust survivors, for example, often recall their reaction to the sun during their ordeal in the camps, noting how "dark" it seemed to them, marking, as one of them put it, "only the beginning of another day, ending—in what?" Its heat could not warm cold souls, already numbed by hunger, disease, exhaustion, fear, despair, and the murder of other family members. Its consoling warmth, in other words, and the certainty of its daily return roused uncommon responses in the context of adversity; this is only one of the signs we must learn to read differently if we are to enter the realms of the Holocaust on its own terms.

A character in one of Aharon Appelfeld's novels, exasperated by language's habit of cluttering the mind with verbal filters against the truth, calls for a "cease-words," and this reflects Appelfeld's own fear that familiar but feeble formulas of solace or evasion would rush in to fill the void of understanding left by the Holocaust. Once I believed that only the literary artist possessed the stylistic and imaginative strategies necessary to gain at least limited access to the ravaged mansion of this catastrophe. But gradually, as these essays will show, I have broadened my vision to include memoirs, diaries and journals, and especially the videotaped testimonies that in their own complex way invite us to interpret the various layers of memory through which the event was experienced by its victims and survivors.

A library of commentary on the Holocaust is available to contemporary audiences in richer abundance than for most other moments in modern history. But because, when fully faced, it exposes a culture of dread, not consolation, we have been slow to greet it candidly. The proliferation of violence in our society, however, may finally have prepared domestic audiences to deal more frankly with the grotesque and gratuitous atrocities that the Germans and their collaborators visited on their victims, especially the Jews. It may sound redundant to speak of *excessive* cruelty among agents already committed to mass murder, but in fact, as episodes mentioned in several of these essays show, such cruelty is exactly what we find in the culture of dread that emerged in Nazi Germany. Only with the greatest of reluctance is one forced to concede what the evidence seems to support: the power of domination, including the power to abuse and kill others, is for some men and women as much a source of *natural* fulfillment and satisfaction as are the expression of charity and love for others. Labeling such deeds "criminal" merely classifies them, without defining or explaining their genesis. The myth that we are restrained in our behavior by the most admirable qualities of our culture was exploded forever by the history of the Third Reich. We are still baffled by the question of why a culture with such admirable qualities ended up finding its most dreadful features even more appealing.

Few today take seriously Hitler's principle that war is the health of the state, but we have been less able to dismiss the other legacy from Nazi Germany—that power is the health of the self. Before their physical oppression began, Germany's victims were deprived of power and ousted from any social bonding that included them as objects of mutual concern. One of my goals in these essays is to evoke the moral dilemma of the human being in a situation of virtual powerlessness, since this is a concrete detail of the Holocaust that continues to elude our imaginative grasp. No one is happy to learn that solicitude is an easy choice only in a privileged milieu, and that as the margins of personal freedom shrink, as they did in the ghettos and camps, human beings are less apt—because less able—to practice the nurturing that a supportive environment makes possible. The image of a self stripped of its moral and physical power that the mirror of the Holocaust reflects back at us is so unflattering, so terrifying in its naked helplessness, that only a few rare artistic talents in this country have had the courage to stare into the abyss and record without flinching what they saw there.

Nearly fifteen years ago, I wrote in "The Americanization of the Holocaust on Stage and Screen" of the then much-acclaimed television drama *Holocaust:*

> The vision which plunges us into the lower abysses of atrocity is not there. We do not know what it was like, in the Warsaw Ghetto and

> elsewhere, to have been reduced to eating dogs, cats, horses, insects, and even, in rare unpublicized instances, human flesh. . . . We see well-groomed and sanitized men and women filing into the gas chamber, but what does this convey of the terror and despair that overwhelmed millions of victims as they recognized the final moment of their degradation and their powerlessness to respond?

Yet viewers here and abroad were grateful for this manageable version of the Holocaust, and ironically, through the universal medium of television, worldwide interest was renewed in a subject that seemed to have dropped from public view. But the sheer brutality of the event was shielded, and the technicolor horror that audiences saw, though mildly disturbing, did not threaten to displace their mental comfort with nightmares.

Today, the phenomenon of *Schindler's List* allows us to measure the inroads that the culture of dread has made on the culture of consolation. Whereas the directors of *Holocaust* found it necessary to include the requisite love interests—Meryl Streep improbably knocking on the gates of a concentration camp to join her deported husband, and a wedding among partisans in the woods—*Schindler's List* decides to focus on the steady fear that drains all feeling when one's daily diet is the ruthless and impulsive cruelty of the murderers. Throughout the film, no one can predict when or why a German will choose to shoot a Jew, until the usual assurances of security granted by familiar versions of human bonding disintegrate before our eyes. With few exceptions, director Steven Spielberg resists the temptation to let old values invade new terror, so that when the women are rescued from Auschwitz and returned to the factory, their men can only stare in disbelief through the windows at their temporary luck: rejoicing is out of the question. Similarly, when most of the children are trucked to their death from the Plaszow labor camp while their mothers are subjected to a "selection," the women can only wail in helpless despair at the children's departure. Spielberg's major achievement has been to eliminate the context of normalcy from the lives of his victims, a context that remains absent until the film's Technicolor return to chronological time in its closing moments. Then we are given a chance for mutual mourning and for solitary meditation on the difference between cinematic vision and the jagged intrusions of history.

Some of the objections raised to *Schindler's List* seem to reveal the very confusions that the essays in *Admitting the Holocaust* seek to address. The complaint that Jews are not sufficiently characterized might be valid were the subject not the Holocaust, which signaled not only physical destruction, but the death of the very idea of the self. Since by training and expectation we still consider individualism to be the most precious legacy of the Western mind, we demand from experience a challenge to personal growth; when that is missing in literature or film, we label it a failure in characterization. But if private suffering is a useful discipline (and this itself may be more myth than truth), mass atrocity is not. The frantic surge of mobs of people from the shelter of their crowded homes into unprotected public spaces and their subsequent journeys to ghettos, work camps, or (via cattle car) a killing center are followed by the camera without benefit of intense scrutiny of inner responses. When asked what they felt during those critical moments then, survivors today reply almost uniformly: "You didn't feel anything; you just wanted to stay alive." *Schindler's List* uses visual effects to create chaos rather than form, and when such chaos invades the frontiers of the mind, it leaves little leisure for an orderly response.

To make the victims in the film more concerned with the question of who you are than with the dilemma of how to *be* would have been to falsify the reality on which the experience of Schindler's Jews was based. Constant noise and confusion, accompanied by a relentless German brutality—a milieu, in other words, that is environmentally determined—replace reflection as the mode of portraying "character," at least for the victims. If this seems insufficient, the fault is not in the film, but in a cultural tradition and mental attitude that demand normal intellectual reactions from abnormal historical situations.

Similar objections have been raised to the film's failure to clarify Schindler's motives for saving the Jews in his factory. Ambiguity seems to be no more native to the American mind than tolerance of obstacles. Among the many notions sabotaged by the Holocaust is the naïve opposition between good and evil or heroes and villains. Although several studies of altruism during this period have tried to isolate reasons why citizen X agreed to hide Jews while citizen Y betrayed them to the Germans, no one has yet analyzed convincingly this elusive mystery of human behavior. *Schindler's List*, like all serious art, invites us to join in the creative process by speculating about the riddle of human nature without expecting simple answers, or perhaps any answers at all.

One of the film's boldest achievements is the attempt to render in depth the commandant of Plaszow, Amon Goeth. A striking feature of Holocaust fiction and drama is the virtual absence of Nazi characters as leading figures: the SS doctors in Rolf Hochhuth's *The Deputy* and William Styron's *Sophie's Choice* are mere allegorical spokesmen for a point of view, while George Steiner's *The Portage to San Cristobal of A.H.* offers Adolf Hitler as a *voice* rather than a full-fledged human being. The real parallel with Goeth is Styron's portrayal in *Sophie's*

Choice of Rudolf Hoess, commandant of Auschwitz. But in trying to humanize Hoess, Styron simply ignores his role as a killer of Jews, pushing his violence to the periphery of the novel, where it is allowed to disturb neither the author nor his readers. Spielberg is not guilty of this evasion. In one of the film's most difficult moments, Goeth's erotic confrontation with his young Jewish housekeeper, the allure of love loses to the appeal of power, reenacting before our eyes the modern paradigm about which I spoke earlier. All viewers may not be satisfied by the version of the Nazi mind projected by the characterization of Goeth, but it has few equals thus far in American literature or film.

To no one's surprise, some relics of Hollywood infiltrate *Schindler's List,* reminding us how powerful a hold the culture of consolation still exerts over the culture of dread. At a time when Schindler would have been fleeing for his life from the approaching Russians, he pauses to deliver his self-disparaging "I should have done more" speech, moralizing his complex conduct and feeding it in easy doses to an anguished and distraught audience. Even more deplorable is his warning to the SS guarding his factory at war's end that they could return home as murderers, or as decent men. That was no longer a choice for such troops in 1945, and even though there is some evidence that the real Schindler uttered those very words, Spielberg was free to reject them for the sake of the artistic logic of his film. His lapse suggests that even he was reluctant to leave his audience on the brink of the moral chaos to which Nazi Germany had led the Western world. He insists on pretending that decency is a salvageable virtue for those so recently conspiring in mass murder, as if the lives of the Jews could suddenly matter more to them than saving their own necks from swift Russian justice. Despite its candid representation of the ordeal of the Jews during World War II, even a blunt film like *Schindler's List* decides to leave us with memories of a healing wound rather than a throbbing scar.

In these essays, I examine some of the reasons why, fifty years after the event, the hope to "cure" continues to seem a better option than the willingness to endure, and try to suggest ways of surmounting the barriers that curb frank encounter. One healthy recent step has been the opening of the Holocaust Memorial Museum in Washington, D.C. Against almost everyone's expectations, it quickly became the most-visited museum in the national capital, which is a tribute not only to the designers, but also to a latent need in the population for a chance to admit the disaster into their consciousness in a gradual but relentless and unsentimental progression toward the truth. The fluidity of human response to catastrophe is certainly evident among those who gathered the remnants of their shattered past and began to build their lives anew after the war. But such laudable adaptation should not be used as a lure to amnesia about the ruins on which their reconstructed future lay. To its credit, the Holocaust Memorial Museum does not try to minimize the misfortune by celebrating the success of its survivors.

Those who managed to stay alive mention their success in fragments of filmed testimony that are the last words museum visitors hear before they leave the shadows of annihilation and return to the bright light of the "normal" world. These witnesses to survival offer us the sanctuary of chronological time to protect ourselves from the utter disintegration of values with which their ordeal afflicts us. But if we have heeded the exhibits that precede this final meeting with the Holocaust's former victims, we understand that enjoying the present while postponing death to the distant future provides a very fragile serenity indeed. Admitting the Holocaust has changed the nature of our fate, since dying in the past—the dying of others, under conditions of atrocity—now shares the stage of consciousness with death in the future. Textbook theories about self-actualization, the intrinsic goodness of the human spirit, moral growth, social progress, and the valuable lessons of history collapse into pretentious evasions of the grim legacies that twentieth-century reality has left us: the Holocaust above all, but only as the chief example of companion forms of mass dying through war, revolution, famine, repression, and genocide. The challenge before us is to rewrite the internal scenario of our lives to include the deaths of others—the victims of our violent time. Through their skirmish with the Holocaust, survivors have been forced to do this, and their testimony becomes a model for our own efforts. As consciousness of private and public dying now shifts between memory and expectation, we are granted a more reliable if more limited invitation to the promises of life.

1994

1

Memory's Time: Chronology and Duration in Holocaust Testimonies

ne of the most striking events in Holocaust studies in recent years has been the proliferation of titles focusing on a single theme: memory. Saul Friedländer's *When Memory Comes* (1979), Pierre Vidal-Nacquet's *The Assassins of Memory* (1985), Charlotte Delbo's last work, *Days and Memory* (1985), Sybil Milton's *In Fitting Memory: The Art and Politics of Holocaust Memorials* (1991), my own *Holocaust Testimonies: The Ruins of Memory* (1991), a collection of essays on the convent controversy at Auschwitz called *Memory Offended* (1991), another set of essays on Elie Wiesel called *Between Memory and Hope* (1990). We also have James Young's study of Holocaust monuments called *The Texture of Memory* (1993), and Geoffrey Hartman's edited volume of Holocaust essays called *The Shapes of Memory* (1993). And I'm sure there are many others. Perhaps this means we have finally begun to enter the second stage of Holocaust response, moving from what we know of the event (the province of historians), to how to remember it, which shifts the responsibility to our own imaginations and what we are prepared to admit there.

If we are to trust the titles, memory itself is not neutral. *Between Memory and Hope* implies one agenda; *The Ruins of Memory*, another. One tries to edify; the other, to plunge us without life preservers into the maelstrom of the disaster. In the end, unlike Dante, the pilgrim of the Holocaust must learn to mistrust all guides, whether they lead us toward closure or not. Unlike history, memory can be a very *private* adventure, and when the Holocaust is its object, it can be a threatening one. Virgil could lead Dante through an Inferno ruled by external order: the consolations of Paradise lighten for the reader the harrowing experience of Hell. Surviving Holocaust victims, in one dimension of their testimonies, disclose an internal *dis*order void of consolations. Dante's pilgrim (and Dante's reader) moves from time to eternity; spatial and spiritual progress accompany each other. But as we shall see, in Holocaust testimony time itself divides, so that memory must contend with the paradox enshrined in Charlotte Delbo's statement "I died in Auschwitz, but no one knows it," or in the words of the former deathcamp inmate who said, "One can be alive after Sobibor without having survived Sobibor."

Sociologist-historian Pierre Vidal-Nacquet calls himself a "memory-man," and has confidence in the power of memory. Philosopher Jean-François Lyotard is not so sure, concerned as he is with what memory "misses" through rejection or suppression. If there is a history of remembering, he argues, there is also a politics of forgetting. Holocaust memory entreats us to accept the destabilizing force of its content, even as traditional memory seeks ways to restore the balance. For example, there is no need to rehearse here the role of the funeral oration in allying grief to memory and celebrating the life which has just ended in death. Such commemorative moments foster communal unity and remembering. But funereal testimony differs from the funeral oration, since it rises not from the spring of chronological time but the tomb of what I call durational time: the one celebrating a life that has just ended in the sadness of death, transcended by a "normal" future for those who remain; the other signifying a death that has not been preceded by a life connected to such an end, followed by a temporal void.

To illustrate these distinctions, let me turn to Abraham Lewin's description in *A Cup of Tears: A Diary of the Warsaw Ghetto* of the funeral in the ghetto of the distinguished Polish Jewish historian Majer Balaban, who in December 1942 died suddenly of a heart attack:

> He was 65 years old at his death [writes Lewin], but looked a lot older. The remains of the Jewish intelligentsia came to pay their last respects to the deceased. God! What a tragic and depressing sight the gathering at the funeral made. Firstly the small number who came. It was plain to see that very few of us are left, just a handful. And secondly: the appearance of the people! What impoverishment, what gloom, what weariness filled their faces! In this small gathering was expressed our total destruction in its most tragic and appalling form, the destruction of the greatest Jewish community in Europe. I was shaken to the core, looking at the extinguished and despairing eyes, and the lined faces and the torn and ragged clothes. Utter annihilation.[1]

We can remember in order to protect continuity or to verify disruption, to conceal or to flay. Both modes prevail in the figuration of the Holocaust, but only when we lapse into confusing the communal ritual of funeral orations with the estranging rite of funereal testimonies. Lewin was honest enough to recognize the difference, and left us a striking instance of how the threat of mass murder had slain an ancient tradition. Lyotard addresses this dilemma when he calls for "an aesthetics of the memory of the forgotten, an anesthetics."[2]

"Here," says Lyotard, "to fight against forgetting means to fight to remember that one forgets as soon as one believes, draws conclusions, and holds for certain. It means to fight against forgetting the precariousness of what has been established, of the reestablished past; it is a fight for the sickness whose recovery is simulated."[3] Simulated recovery belongs to the realm of chronological time. In the realm of durational time, no one recovers because nothing is recovered, only uncovered and then re-covered, buried again beneath the fruitless struggle to expose "the way it was." Holocaust memory cannot be used to certify belief, establish closure, or achieve certainty. Hence chronological time is needed to intrude on this memory by those who insist on rescuing belief, closure, and certainty from testimonies about the disaster. Durational time resists and undermines this effort.

For Lyotard, Holocaust past is "a past located this side of the forgotten, much closer to the present moment than any past, at the same time that it is incapable of being solicited by voluntary and conscious memory,"[4] and this is what I call duration, which exists *this side* of the forgotten, not to be dredged from memory because it is always, has always been there—an always-present past that in testimony becomes a presented past, for the witness more precisely a re-presented past, and then, in narrative forms other than testimony, a represented past. What is lost during the transitions?

How can written history, which began in the form of "annals," avoid chronology? When the annalist becomes an analyst, he also becomes a self-conscious servant to time. Do testimonies tell us about what precedes or dominates one's present life? Yes—and no. The duration of Holocaust time, which is a constantly *re*-experienced time, threatens the chronology of experienced time. It leaps out of chronology, establishing its own momentum, or fixation. Testimony may *appear* chronological to the auditor or audience, but the narrator who is a mental witness rather than a temporal one is "out of time" as she tells her story. I as interviewer have the strange feeling that I am "in" time as I listen, yet she is "out" of time as she speaks. I try to enter her *mental* space, not her temporal or chronological one. Lyotard calls it the retrieval of a time that is lost "because it has not had place and time in the psychic apparatus." He speaks of a moment "where the present is the past and the past is always presence,"[5] while at the same time admitting the unsuitability of the very terminology he uses.

But this is often the case in Holocaust discourse; we are led astray, or baffled by the lack of a language to confront the difference between the chronological current, which flows until we channel it between the permanent banks of historical narrative, and durational persistence, which cannot overflow the blocked reservoir of its own moment and hence never enters what we call the stream of time. Or at least not until I alter it by trying to write about it as if it had a before, a during, and an after, because most writing cannot exist without the temporal succession that violates the uniquely imprisoned persistence of the Holocaust event in the memory of its witnesses. The most successful artistic attempt I know to overcome these obstacles in narrative prose is Jorge Semprun's novel *The Long Voyage* (in which the narrator "remembers" his future, a logical impossibility, but an artistic reality nonetheless).

To illustrate what I mean by the uniquely imprisoned persistence of a Holocaust event in a witness's memory, memory's time or more precisely durational time, let me turn to the testimony of Anna S. (not her real name). Anna married at eighteen because her parents thought that a husband might help her better to survive in the harsh conditions under Nazi rule. But after the Germans invaded their home one night and shot her father and brother—she, her husband, and her mother managed to hide or flee—husband and wife separated, she and her mother sleeping in barns and he leaving with his brothers for the forest. (She learned much later that they were captured, tortured, and executed by the Gestapo.) All this time Anna had been pregnant, and as she neared her term, her mother insisted that they find quarters in a nearby ghetto. Here Anna gives birth, while she is suffering from typhus, in a cold, tiny room where, as she tells us, she is sleeping in a *crib* (others are sleeping beneath it) because that is the only available bed. When she goes into labor, her mother rushes out to find a neighbor who had promised to help with the delivery. While she is gone, the baby is born. The testimonial moment begins:

ANNA And I was burning up with typhus, and then—the baby came out.

INTERVIEWER Stillborn?

ANNA Hm?

INTERVIEWER Stillborn.

ANNA No.

INTERVIEWER Alive?

ANNA It was alive. It was alive.

INTERVIEWER Was it a boy or a girl?

ANNA Boy. Yes, he was born alive, and we didn't have a—doctor's no question—we didn't have a midwife to take the baby.

At this point she tells of her mother running out in search of the neighbor, returning just after the birth had occurred. Then she resumes (and note her *utter* inability to reconstruct these moments in any kind of chronological order):

ANNA So finally, finally, I don't know how long it took, I was unconscious. I was like, I knew it happens something now, I couldn't, I don't know, I couldn't respond to anything. And then finally she came in, I don't know how long she was there, I heard (she imitates choking sounds) he was choking, he was choking [whispering] and he died, he died, when she came he was dead already. [. . .] After a minute I woke up and gave a look. It was a beautiful boy, a beautiful boy. [. . .]

INTERVIEWER So you were alone all that time?

ANNA Yes. I was alone, and next to mine bed, mine crib, was a man dying, and I opened my eyes and I looked at him and he was dying—and I fall asleep.

Holocaust testimony enacts a resistance against the efforts of time to erase experience without a trace: what Lyotard calls "time lost yet always there, a revelation that never reveals itself but remains there, a misery."[6] For how could Anna's words reveal to us what that moment must have been for her, what it remains, not in conventional memory, since she is obviously not "remembering" a forgotten moment? Her uttering of details, naming the unthinkable, her enactment for us of the truth that at the same time this self is *not* that self—she has remarried, and had a new family—and *is* that self, our difficulty in finding a familiar context or designation for what she describes, our inability to detect through a sequence of events the presence of a guilty agent somewhere in the obscurity of the past—all of these together are gathered in a cornucopia of diverse causes, which in the end paralyze our capacity to judge, evaluate, or perhaps even respond. We are confined, consumed by the moment of the narrative which is not a moment in sequential time, mesmerized by duration until chronology disappears from consciousness. These segments of testimony enable us to experience the effects of Holocaust duration as no other form of expression can.

Hearing testimony, we are in the presence of a past that has not been and cannot be effaced, a moment re-presented to us rather than represented, since, as Lyotard insists, only that which has been inscribed or represented (in word or image or form) can be forgotten. Such testimony disrupts chronology, where we are safely situated, and drags us into the menacing realm of duration, where nothing can shield us from this unpresented or never-before-presented rehearsal of inversion, where death replaces birth as the "normal" course of events, only to have narrative restore the *illusion* of chronology. By this I mean that circumstances determine the baby's death before it is born, though chronology prompts us to hope that maybe it will be rescued. In describing the birth of her baby that is already doomed, Anna lures us into her own experience of duration, and for this instant, anyway, we escape what Lyotard calls "the crude forgetting of the unforgettable secret."[7] When, some minutes after this part of her testimony, Anna leans forward and whispers into the darkness beyond the cameras, "I never told this story to no one before," she is marking us with the scars of that hitherto undisclosed mystery of her ever-present past, which until now for us has been a never-present past. Where do we then fit it into the chronology of humanistic endeavor, of healing old wounds, of reconciliation, of redemption? Time as chronology does not and cannot heal the wounds of time as duration.

Chronology anticipates something. Duration anticipates nothing. Birth is a culmination of developing time. Its features are growth and nurture, expectation, the organization of various biological, chemical, and physical forces into life. Like history, it represents a gestation from the past in the service of the future. The disaster of the Holocaust violates such chronology in nature, and this moment sheds a glaring light on the reversal and disruption accompanying the disaster. The victim is powerless to prevent the disintegration of what for nine months has been integrating within her. What system of belief can we build from the debris of this testimony? I suppose I asked, "And what happened next?" after this moment of narrative (since I was the interviewer), but what answer could I have expected? *Everything* happened afterward—and nothing. As Charlotte Delbo implied in the title for the trilogy of her Auschwitz memoirs, *Auschwitz et après,* there was an Auschwitz, and there was an afterward, and unless you understand that the two terms do not represent a chronology, you cannot begin to enter the abyss of the place we call Auschwitz. Such testimony *serves* nothing; it can only *pre*serve.

What else can it do? It can liberate the inaudible from the silence that insulates it from our ears *because* it is unbearable; testimony like the one we have been examining helps to make certain silences audible, creating the paradox of the audible silence. Unfortunately, we are still left with the burdensome legacy of the *in*audible silences, echoing from the inner walls of the gas chambers, whose text we will never hear. They will remain inaccessible and hence inaudible forever; we will have to make do with the fragments of audible silences like this one.

If we cannot find an aesthetics for Auschwitz, we must be content with what Lyotard calls an "anesthetics." Amnesia has been the comfortable stance of those disinclined to traffic with the threats of audible silence. As anesthesia is the medical means for making one insensible, anesthetics deals with the art of the insensible (and in a related way, with the non-sensible), plunges us into the non-sense (not the nonsense) of the disaster, reminds us that no ordinary feelings will make us sensitive to the appeal of such an unprecedented catastrophe. If art is concerned with the creation of beautiful forms, Holocaust testimony, and perhaps Holocaust art as well, deals with the creation of "malforms," though we may not yet have arrived at recognizing the legitimacy of this undertaking—to say nothing of the word itself. Birth-as-death makes the unnatural "natural," and though we resist the logic of this conjunction, it pursues us and persists through the duration of such testimonial evidence as the one we have just heard.

Life-as-death was the "natural" mode of being in places like Auschwitz, just as dying, for the Jews, was the goal and purpose of living under the Nazi regime. Perhaps we need to add to the vocabulary of atrocity a term like "maldeath," in order to avoid confusing the episode of the baby's birth-as-death with forms of normal dying. There is only one way to present such death, but many ways to represent it. If we speak of the dignity of dying, of the triumph of the human spirit in a place like Auschwitz, we risk a bizarre effort to transform mass suffocation, with its twisted, entwined bodies stained by human waste, into an instance of what Lyotard calls "beautiful death." "The only way you can make a 'beautiful death' out of 'Auschwitz' death," he says, "is by means of a rhetoric."[8] This may be a major stylistic achievement, but it has little to do with the truth. The idea of a "beautiful death" invoked by words like "sacrifice" and "martyrdom" awakens the consolations of chronological time, but the price one pays is fidelity to the durational moment: that is, the last gasp of life, followed by the literal annihilation of the remains, until no trace is left, in memory or fact, of what was once a human being. The language of chronology seeks to rescue the victim from that mute fate.

Lyotard maintains that Holocaust art "does not say the unsayable, but says that it cannot say it."[9] Is this true of testimony too? Perhaps not. Surviving victims bear witness to the impossibilities of their lives then; we tend to translate them into possibilities by easing them into chronological time as wounds to be healed, insults to be paid for, pains to be forgotten, deaths to be transcended or redeemed. Martyrdom or the notion of dying with dignity reestablishes the communal "we" that makes the victim's death in Auschwitz a "shared" experience with the future and the past, and with prototypes from history. Such referents convert death in the gas chamber into a "norm" of dying that makes it accessible and even "acceptable" to the generations that have to live with the memory of this atrocity. Martyrdom and dying with dignity not only furnish an unthreatening form of discourse, but restore the individual victims to the stream of humanity in time and thus rescue them from the anonymity to which the Nazis had consigned them. The "beautiful death," as Lyotard describes it, exchanges "the finite for the infinite"; it becomes an affirmation of the formula *"Die in order not to die."*[10]

But Auschwitz signifies a splitting, not a continuity; speculative discourse about it seeks to restore—or to discover—a linkage. Nazism leaves a silence behind it, Lyotard argues, because its premises have not been "disproved"—it has been beaten down by sheer force. It has not been outargued; perhaps that is one reason why deniers today can maintain that the Holocaust never happened. The Nazis and the Jews had no stakes in common, but even harder to accept is the truth that the fate of the Jews and our own chronological world view have no stakes in common. I am tempted to add that the fate of the Jews and the timebound historical imagination may have no stakes in common either. Just as the Nazis and the Jews inhabited different universes of discourse (it took the victims a long time to admit this; when they realized it, it was for most of them too late), so we who come "afterward" inhabit a different universe of discourse from the one that the victims lived and died in. They were nullified, not sacrificed; murdered, not martyred. Just as there was no common idiom between the victims and the killers, there is no common idiom between the killed and ourselves. The Jews were forbidden by Nazi indifference and contempt to thrive as a Jewish community despite oppression, or to achieve in the ghettos and deathcamps the heroic response that some commentators would like to impose on them. Only through the invention of a mythic narrative "afterward" can we reconstruct an idiom to change their death from a "forgettable" (because unbearable) occasion into a memorable one. But this only grafts speech onto silence; it does not illuminate the silence. "Those silences," insists Lyotard, "signal the interruption of the *Selbst* [self], its splitting apart."[11] And this splitting of the self is a major and permanent legacy of the event we call the Holocaust.

Durational time does not believe in or allow for any sort of foreclosure toward the past: *its* foreclosure expels the future. Anna S. utters (or whispers) the inaudible silence that we strain to hear, making strenuous demands on us as we listen. Her words are unforgiving and unforgivable, insisting on the privacy of her narrated ordeal, isolating it from the chronology that might change it into a communal experience, sanctified by analogy with a precedented grief. Other women, after all, lose children during the birth process. Duration mandates and ensures the inviolability of her unprecedented violation. Seminal moments in the testimonies are invariably death moments; given the nature of this catastrophe, we should not be surprised. All the witness can do is to release them from their hermetic grave into living consciousness, where they remind us of the limitations of the sanctuary we call chronological time, our normal refuge from the unmediated misery of what we have just encountered.

I would like to conclude with a brief confrontation with another testimonial moment, when chronological and durational time literally intersect in the telling (or speaking), as the witness describes his reentry into the flow of chronological time even as durational time continues to flood his psychic life. The tension of this unendurable struggle is visible in his anguished expression as he describes it. And what is our role? Once again, Lyotard provides us with an aphoristic basis for our response: "[T]hought," he says, "cannot actualize, act out, the return of the disappeared but merely watch (over) the Forgotten so that it remains unforgettable."[12] This surviving victim, as we see, needs no such caution.

The chronology of George S.'s life took him to America after the war, where he tried to settle down into a normal existence. But because his life had been *unsettled* in Europe during the war, he was left with a legacy of unresolved conflict that emerges starkly from this moment in his testimony. After more than five hours of recapitulating the destruction of all but one member of his extensive family, he abruptly starts to speak of his fear of marriage and beginning a family following such ruin. "A home," he defines, preparing us for his narrative of how duration invades chronology, "is something you lose." Nevertheless, he finds a job, marries, and begins a family. He has certainly earned an unimpeded future, but we are naive to expect its fruition:

> We bought a home, started to establish ourselves, and things were getting better—but I had some drawbacks. I was working hard, trying to forget myself, forgetting the past, but it came back to me like a recorder in my head. After we got married, for the longest time—we were already then in a family way, and things were looking up to me. During the day I was working hard, and studying, and trying to get ahead and establish myself—and at night, I was fighting the Germans, really fighting. And the SS were after me all the time, and I was trying to save my mother and my sister. I was jumping off from

> building to building and they were shooting at me, and each time the bullet went through my heart. And I was sitting up not knowing at night in my bed, and I was screaming, you know it was hard on [my wife]. It must have been hard on her, and we didn't know how to handle it. I didn't know how to handle it. I was making believe . . . forgetting about it and going on during the day doing my things. And at night, she was calming me down, she says "It's OK, it's OK, you're here, don't worry about it," and I was waking up and screaming at night and each time the bullet went through my heart.[13]

He returned to work, but his left arm began to hurt, he had difficulty driving the car, and one day a fellow worker rushed him to the hospital with the symptoms of a heart attack. The doctors found nothing physically wrong with him.

He could hardly be more explicit about the two currents that buffet him: "During the day I was working hard, and studying, and trying to get ahead and establish myself—and at night, I was fighting the Germans, really fighting." He admits that in his recurrent nightmare he is trying to save his mother and sister (both gassed at Auschwitz), a futile struggle, of course, but one that persists as the very essence of durational time. Is the bullet through his heart an unconscious (if unsuccessful) effort to banish that time and renew his life by liberating it for the future? Or is it a violent reminder that durational time will continue to assert its disintegrating claim over his bid for integration? Death, after all, was his destiny too in the camps; his survival, in a profound and significant sense, is an *evasion* of his intended fate. He lives, in Charlotte Delbo's indelible words, *en sursis*—under reprieve. If I may be permitted a paradox, the self-contained "chronology" of durational time normally ends in some form of atrocity, usually death. The bullet through George S's heart reminds us that *every* deathcamp survivor has violated the hermetic "chronology" of durational time simply by staying alive. The relief of *true* chronological time must contend with the grim logic of durational time, which during the Holocaust demanded his death and afterward refused to verify or validate his survival. It is clear from these testimonies that durational time relentlessly stalks the memory of the witness, imprinting there moments immune to the ebb and flow of chronological time. No public ritual can ease the sting of such private recall, which persists outside the frame of consolation or closure. The lingering of durational time distinguishes it from the evanescent flow of chronological time, with its before, during, and after. The latter not only tempts, but entreats and *enables* us to forget the unforgettable. I have great fear—though I hope I'm wrong—that decades hence, these testimonies may be moldering in their archives, victims, like the witnesses whose forlorn stories they contain, of the welcome amnesia of chronological time. But meanwhile, they challenge us to keep watch, in Maurice Blanchot's memorable words, over the absent meaning that continues to distress us all.

1992

2

Beyond Theodicy: Jewish Victims and the Holocaust

he situations of Job and of Jesus have created a pattern for thinking about suffering that continues to inspire the human imagination up to the present time. Two of the most famous lines from the Gospels, spoken at Jesus's moment of ultimate agony—"My God, my God, why hast thou forsaken me?" and "Forgive them, Father, they know not what they do"—are addressed to a divinity in whose existence the speaker has total faith. The momentary despair of the first query and the compassion of the second comment leave faith in divine power and love intact, even while admitting the human limited ability to understand their operation.

The ordeal of Job—or test, as some would have it—similarly addresses the problem of the difficulty in understanding the operation of divine power and love. Like the story of Jesus, the story of Job has become a parable of the mystery of human suffering. Job, however, refuses to accept the justness of his suffering, insisting on his own righteousness and challenging his God to justify his ways to humanity, while refusing to embrace his wife's advice to "Curse God and die!" He attempts verbal battle with his God (itself an expression of his faith), and is indeed rewarded by the rare experience of a direct address from the Divine Voice. In the end, we are asked to believe, Job's adversity strengthened his moral will and spiritual integrity. Like Jesus's (though with fewer theological implications), his suffering was a form of martyrdom. Today, both figures remain archetypal examples of the value of suffering for the growth of the human spirit.

They also remain virtually useless in helping us to understand the Holocaust experience, though I realize this statement may sound radical, unorthodox, or threatening to many readers. Nonetheless, I am convinced that trying to "read" the Holocaust through the values implicit in the stories of Jesus and Job leads us from the uncharted waters of that atrocity back into the safe channels of a sheltered world, where harbors are well-lighted and voyagers disembark feeling that the human journey has had a purpose. That voyage has always described evil as a negation or violation of good. Few, apparently, were prepared to concede (few still are) that in the other world of the Holocaust, what we consider evil was for the Nazis an *expression* of good, supported by a political and moral value system totally alien to our orthodox minds.

If the Holocaust has taught us much about the vulnerability of the human body, it has taught us more, perhaps, about the fragility of words themselves. Words not only describe the essence of the Holocaust experience for those who planned, endured, or survived it; they also manipulate, alter, and revise it. Words infuse, diffuse, and confuse meaning; we follow the flickering torch of vocabulary sustained by the security of earlier verbal illuminations. When Viktor Frankl, reflecting on his time in Auschwitz, quotes Nietzsche's "That which does not kill [*umbringt*] me, makes me stronger," he imposes on the innocent reader several verbal assumptions that are far from unquestionable. "Kill" implies a form of dying (as well as a form of execution) that stirs the imagination with vague or precise associations, depending on the context one provides. The Nietzschean allusion furnishes a context of "will," encouraging the association with heroic individualism that was characteristic of much thought in the nineteenth century. Although Frankl does not invite this response, the wary Holocaust voyager must ask whether that which kills my mother and father (or wife, or children, or neighbors), but not me, also makes me stronger. How are we to interpret that word "stronger" *(stärker)* when we confront it with the challenge of surviving daily or weekly or monthly selections for the gas chambers, in a deathcamp like Auschwitz-Birkenau? In other words, when the vocabulary used to describe this event is neutralized by its cultural context (here Nietzsche, will, heroism), shorn of the specific forms of terror that the victims actually faced (gas, fire, frost, heat, starvation, thirst, beatings), we begin to recognize how easily language can be used to *betray* reality, to reshape (or distort) it, indeed, to summon up any attitude we wish, *despite* the inhuman conditions that inspired the attempt.

The Holocaust is an expression of a particular atrocity, not of prior religious or historical moments of suffering. The failure or reluctance to admit this leads to much confusion and misunderstanding. The crucifixion of Jesus and the torment of Job are reported to us in Scriptures as part of a complex divine vision of an unfolding universe, which includes (but not exclusively) humankind. The destruction of European Jewry was not part of a divine plan, but a human one. Some day an eschatology may evolve around it, but that will transform the event into something it was not in order to make it still more acceptable to the human imagination. Even the current disproportionate emphasis on *resistance* during the Holocaust confirms our view of it as a secular rather than a divine event. This emphasis seeks to graft a heroic face onto the catastrophe though the scope of the loss hardly vindicates such an effort. The exodus from Europe and the ghettoes to the deathcamps was not led by a Moses capable of miracles, conversant with the Divine Voice; the only Promised Land awaiting the travelers was not a spacious desert, but infinite misery in a little room.

By their conscious design to remove all Jews from the ranks of the human, the Nazis intended to deprive them of the "privilege" of verbal categories that sustain order for the rest of us; Jesus and Job in their times did not have to face similar exiles from the word. Jesus was "tried" and Job was "tested" by divine permission or will; but the Jews were not. The spiritual resonance of words like "justice," "injustice," "suffering," "martyrdom," and "heroism" depends on our freedom to appeal to an external authority or established tradition for their validation. Jewish suffering ceased to be exemplary when the circumstances imposed on the Jews by their persecutors (not *ever* by their own choice, or will or indifference) cut them off from appeal—for sympathy, status, or justice. Such a situation must have been as unbearable for them as it is for us—an exile not *of* the word, but *from* the word. The spiritual desolation of such an existence is highlighted by Viktor Frankl's desperate if misguided avowal: "If there is a meaning in life at all, then there must be a meaning in suffering." But he gives away the game when he concludes that "without suffering and death, human life cannot be complete."[1]

The irreducibly terrible point of Auschwitz and other killing centers was precisely that in those places the natural completion of human life was violently, viciously, ruthlessly, and senselessly aborted. If, for example, martyrdom seeks to establish a voluntary link between how one lives and how one dies, involuntary suffocation in gas chambers rudely severs that potential connection. By the mass, anonymous manner of the executions, the killers deliberately sought to defeat martyrdom. Everything the Nazis did conspired to deprive victims of their dignity: they lied, mocked, humiliated, starved, beat, refused to differentiate by eliminating names and substituting numbers—truly, the murder of the word as well as the man and woman. They rejoiced in the sundering of the deathbound from a language of appeal, as well as from each other. Afterward, however, catastrophe is what we make of it, not what it was, and if we try to restore to mute atrocity a voice of appeal in order to make it bearable to ourselves—who can condemn this honest search for a suffering more resilient than it appears at first to have been?

But can we afford uncritically to applaud such an effort? I think not, chiefly because it is so inconsistent with the testimonies of so many of those who endured the atrocities we prefer to call suffering. For the past three years I have been watching videotaped interviews with Holocaust survivors, interviews now numbering nearly 300 and ranging in length from forty-five minutes to more than six hours. For most of these witnesses, liberation brought only limited relief, and survival qualified joy, while a word like "suffering" exposes the poverty of conventional vocabulary to evoke the ordeals they describe. The tales they tell paralyze the will as well as the word. Unintentionally, they ridicule our naïve notions of choice, heroism, and resistance, those verbal creatures of the human imagination that shrivel and disappear in their other world of the Holocaust, where the inhuman imagination holds sway.

In Auschwitz and the other camps, the Germans violated the principle of creation on which our spiritual universe is based and substituted for it a universe of destruction whose "morality"—and here again we stumble over the absence of an adequate vocabulary—excluded the kind of access to faith and divinity available to Jesus on the cross and Job in his misery. This universe of destruction undermined fundamental concepts that normally nurture human consciousness: tragedy, personal destiny, the discipline of private suffering. And it did this by eliminating the possibility, the very idea of "personal" or "destiny." Donne's "No man is an island" was inverted and perverted to read "every man is discredited by the death of someone else." This is what I mean when I say that Jewish anguish in the Holocaust could rarely be exemplary: just as one survival was usually paid for by someone else's death (for example, "I eat = you starve," or "you are selected = I am spared"), someone else's execution tainted the life of those who survived, especially if the victims were members of your own family whom you were unable to help.

The notion of being "unable to help" is one of the most difficult to grasp for those outside the Holocaust universe, because the universe of creation we normally inhabit is based on principles of shared responsibility, love, and mutual concern. These, of course, existed in the camps, but they were intrusions and not expressions of the Nazi universe of destruction. They were sporadic gestures rather than coherent and effective bulwarks against extermination. Nazi evil not only subverted good as we know it; the forms it took poisoned the possibility of a redemptive suffering to counteract the moral paralysis it generated. An episode like the following, one among hundreds of related examples, mocks our efforts to incorporate some intelligible and adaptable view of suffering into our post-Holocaust systems of faith.

In the moral universe of world religions, parents care for children and children care for parents. Spiritual life has always been centered on mutual family concern. To behave otherwise would violate principles of good and evil that govern existence in civilized, reverent societies. Consider, then, the situation of Abraham P. The moment he arrived at Auschwitz on a transport from Hungary with his parents and three brothers remains engraved on his memory, partly because subsequent events proved how unprepared he was for the ordeal. "That day so many things happened to us," he says. "We really couldn't sort them out and I'm still trying to sort them out [nearly forty years later]." His parents and youngest brother are sent to the left, toward death (though of course he didn't know this at the time), while he, an older brother, and a younger brother are sent to the right. Abraham P. recalls: "I told my little kid brother, I said to him *'Solly, gey tsu Tate un Mame* [go to Poppa and Momma].' And like a little kid, he followed—he did. Little did I know that I sent him to the crematorium. I am—I feel like I killed him. . . . I've been thinking whether he reached my mother and father, and that he *did* reach my mother and father. He probably told them, he said, *'Avrum, hot mir gezugt dos ikh zol geyn mit aykh* [Abraham said I should go with you].' I wonder what my mother and father were thinking, especially when they were all . . . when they all went into the crematorium [the gas chamber]. I can't get it out of my head. It hurts me, it bothers me, and I don't know what to do."[2]

This is the "phase of anguish" that, according to Primo Levi, so many survivors must face after their "liberation." To call it a form of suffering that strengthens the spirit, however, would be a travesty of language and an insult to the gravity of the victim's dilemma. Family solicitude led to—indeed, in this instance *ensured*—the doom of Abraham P.'s younger brother. Ironically, *he* feels guilty, while after the war the guilty proclaimed their innocence, or kept silent. His response has nothing to do with the human frailties or natural upheavals that are part of the imperfect conditions of our existence. And who, without fearing the charge of blasphemy, would call Auschwitz "part of the imperfect conditions of our existence"? But how else shall we designate it? When Abraham P. exclaims, "It hurts me, it bothers me, and I don't know what to do," he utters a *cri de coeur* that echoes the chagrin of thousands like himself, and his story begins to merge with parable. His parable, however, darkens the human spirit instead of illuminating it, because it betrays the limitations of all pre-Holocaust spiritual vocabulary when it is applied to that event. He would not think of charging God with injustice, or asking God for forgiveness, because for him Auschwitz is palpably not a problem of theodicy, but of *secular* evil alone. But since only the expression "to the left" remains in his memory to represent the agents of atrocity, he ends up implicating himself in the destruction process, needing to gain forgiveness for, to ask forgiveness from himself. In accepting partial responsibility for a result that he actually shares no moral blame for at all, he relies on the very system of values that the Holocaust temporarily negated. But his dissatisfaction with himself confirms the desperate futility of such reliance.

Instead of examining the relevance of the Holocaust to theodicy, perhaps we need to reexamine the relevance of theodicy to the Holocaust. If we are to believe the testimony of several hundred survivors, the connection is slim indeed. The one theme that weaves through these testimonies like a scarlet thread is the utter irreconcilability of the deathcamp experience with any prior consoling system of values. Life goes on, of course, and innumerable survivors return to the faith of their (murdered) fathers and mothers. But whether from habit, need, or conviction, the return does not solve the not-so-sacred parable of Abraham P. His dilemma remains. Although the present anguish of these witnesses for a past they could do nothing to alter or prevent seems endless, it teaches us nothing about developing an enduring attitude toward suffering that might eventually lead toward some form of healing, forgiveness, or redemption. The atrocities they beheld or endured were beyond suffering, as they were beyond the framework of conventional theodicy. I mean this not as an indictment, but as a melancholy truth that we must face. The Holocaust was a kind of physical and spiritual amputation, leaving parts of the self intact, requiring others to be mended with artificial "limbs." They are serviceable, ingenious, efficient, even sources of pleasure and joy. They sustain life, but they do not and cannot replace what has been lopped off. The twinges that surge through the joint where the remaining flesh meets the prosthesis reminds us of the price we have paid, and must continue to pay, for the "event" we call the Holocaust.

1989

3

A Tainted Legacy: Remembering the Warsaw Ghetto

oes Holocaust "remembrance" have redemptive power? Although this assumption has nurtured an extensive commentary on the Holocaust, it is at odds, linguistically and ultimately factually, with the reality of the survivors' memory. In framing the Holocaust through the lens of heroic rhetoric, Holocaust chroniclers exhibit their own discomfort with the facts left to us by Holocaust victims, dead and alive, and reveal the inadequacy of our language in the face of what there is to tell.

When the German administrator of the Warsaw ghetto told Adam Czerniaków, president of the Warsaw Jewish Council, that starting on July 22, 1942, he wanted 6,000 Jews daily for "resettlement" to the east, Czerniaków, undeceived, decided that he did not wish to preside over the destruction of Warsaw Jewry: the next day, he committed suicide. This is a piece of incontestable documentary evidence, but like much other evidence of this sort, it offers us no guidance.

How shall we, generations later, judge this act? Does it represent courageous defiance or a withdrawal from responsibility? Czerniaków, after all, was the leader of his community, and his voice might have summoned his fellow Jews to some act of public resistance. Decades after the war, Marek Edelman, one of the few surviving leaders of the Warsaw ghetto uprising the following spring, criticized Czerniaków for failing to make a public declaration of the truth—that the Jews were about to be murdered. "One should die only after having called other people into struggle," Edelman charged. He and his dead friends reproach Czerniaków "for having made his death his own private business."[1]

But as Yisrael Gutman, also a member of the Jewish Fighting Organization during the uprising and today a distinguished Israeli historian and director of research at Yad Vashem in Jerusalem, points out in defense of Czerniaków, at the time "even the various underground factions were unable to concur in an appraisal of the situation, address the masses of Jews with a common appeal, or call for resistance as a means of response."[2] We also know from Czerniaków's diary that only two days before the deportations were to begin, he went to half a dozen Gestapo, SS, and civilian officials seeking confirmation of the persistent rumors that a massive "resettlement" operation was about to start, and to a man they denied the rumors as utter nonsense.

Lies and deception were integral parts of the German strategy; Czerniaków's mistake was to believe what he needed and wanted to believe, but he can hardly be blamed for that, since he shared that weakness—or naïveté—with most of his fellow Jews in the ghetto. His death must have been an admission to himself that he had been a tool of the Germans all along, and a sign of his refusal to accept that role any longer. Since there was no organized resistance in the ghetto in July 1942, and little possibility of any appearing, Czerniaków seemed to have had little choice. But this does not make his final act any easier to assess so many decades later.

Edelman tells another story: when the Germans entered the ground floor of the children's hospital in the Warsaw ghetto to round up the young patients for deportation to Treblinka, on the upper floor Adina Blady Szwajger, a Jewish doctor, was busy poisoning the sick children to "rescue" them from that doom. "She saved these children from the gas chamber," says Edelman. "People thought she was a hero."[3] This is perhaps the most bizarre definition of heroic behavior we will ever encounter. The issue is not whether the designation is correct; the issue—here, as in the case of Czerniaków's suicide—is the poverty of traditional moral vocabulary when we address the subject of human conduct during the destruction of European Jewry.

Much writing about the Holocaust, including some works of history, illustrates the failure of language to grasp the thoroughly disruptive, not to say disintegrative impact, of that event on familiar value systems. Marek Edelman goes so far as to call the Warsaw ghetto uprising "undramatic," because for him dramatic action depended on choice, on making a decision. The fate of every Jew in the Warsaw ghetto, including those who took part in the uprising, was predetermined by the Germans: it had already been established that they were all to be killed; any exceptions (and there were very few) would be owing to chance and luck, not choice. As a cardiologist (after the war), Edelman knew that doctors often made decisions that might save a patient's life. But in the Warsaw ghetto, he concedes, "It was always death that was at stake, not life."[4] The victims, especially those who joined the uprising, knew that they were supposed to die. If we raise that awareness to a public level, it means that historians of the period must accept the fact that they are writing of a moment in time without a future. The challenge to us is to try to suspend the privilege of having a future in order to enter sympathetically into the daily ordeal of human beings who could not share it.

Fortunately, we have the testimony of survivors and victims to help us make this transition, but we must be prepared to face a reality that cannot be normalized or sanitized by Romantic references to heroes and martyrs. Most victims did not see themselves that way. Listen to the voice of Abraham Lewin, whose recently translated *A Cup of Tears: A Diary of the Warsaw Ghetto* is one of the most important sources providing us with an account of the daily struggle to stay alive. Lewin writes:

> The proportions of life and death have radically changed. Times were, when life occupied the primary place, when it was the main and central concern, while death was a side phenomenon, secondary to life, its termination. Nowadays death rules in all its majesty; while life hardly glows under a thick layer of ashes. Even this faint glow of life is feeble, miserable and weak, poor, devoid of any free breath, deprived of any spark of spiritual content. The very soul, both in the individual and in the community, seems to have starved and perished, to have dulled and atrophied There remain only the needs of the body; and it leads merely an organic-physiological existence.[5]

Those of us familiar with descriptions of the Warsaw ghetto as a model for heroic resistance and the resolute will to survive may have difficulty reconciling Lewin's desperate portrait with that more congenial version. The conflict leads us to consider the two planes on which the event we call the Holocaust takes place in human memory—the historical and the rhetorical, the way it was and its verbal reformation, or deformation, by later commentators.

Since the Warsaw ghetto has become the emblem of Jewish resistance for many of those commentators we need to balance the attitude based on a rhetoric of heroism with the testimony from those who were there. Probably the most important witness, in terms of the archive of documents he collected and buried, was the historian Emmanuel Ringelblum. After the war, ten cases and two milk cans of his records, diaries, journals, and historical commentary were discovered in the ruins of the ghetto. They represent an invaluable legacy of a man who tirelessly inspired a staff of writers—Abraham Lewin was one of them—to preserve for history the narrative of their people's ruin. This was not only Ringelblum's life work; it was also his death work, since when he had an opportunity after the uprising to be smuggled out of the country by the Polish underground, he refused. He preferred to continue writing in hiding until in March 1944, the

Gestapo discovered him and his family and thirty-five others and executed them all.

One of the most famous passages in Ringelblum's notes is the entry for October 15, 1942, about a month after the great deportation was temporarily suspended. Writing as historian rather than psychologist, Ringelblum raises an issue that continues to trouble many students of the Holocaust today:

> Why didn't we resist when they began to resettle 300,000 Jews from Warsaw? Why did we allow ourselves to be led like sheep to the slaughter? Why did everything come so easy to the enemy? Why didn't the hangmen suffer a single casualty? Why could 50 SS men (some people say even fewer), with the help of a division of some 200 Ukrainian guards and an equal number of [Latvians] carry the operation out so smoothly?[6]

Ringelblum neglects to mention at this point the role of the Jewish police in the roundups, although he was certainly aware of it. Promised exemption for themselves and the members of their families, the Jewish police in the ghetto played an active and sometimes a brutal part in helping the Germans with their plans. They did it not out of hatred, but fear—a natural, if not a particularly commendable, response. Threatened by similar dangers, victims from other nations behaved in the same way. It was a human, not a Jewish, reaction, made ironic by the fact that in the closing days of the deportations, the Jewish police and their families were themselves shipped off to Treblinka. Ringelblum's rhetorical questions reflect the despair of a man who has witnessed the disappearance of hundreds of thousands of his people. But they are also a trifle naïve, and if we continue to ask them today, when we know the total has risen to the millions, we share in that naïveté.

The questions Ringelblum raises are troublesome and remind us that documentary evidence is only the beginning of our inquiry, since it provokes the need for interpretations that threaten our comfortable belief in the redeeming power of memory. Remembering the Holocaust is the most unredemptive task one can possibly imagine. Ringelblum clearly knew that many hungry victims went voluntarily to the trains in Warsaw because the Germans promised bread and marmalade to those who complied with their order. Pleas to the human spirit, to brotherhood and cooperation, to say nothing of physical resistance, usually fall on deaf ears when those appealed to are driven to the verge of death by hunger. And even when, some months later, the Jewish Fighting Organization was finally molded into a resistance group in the ghetto, it numbered only several hundred—estimates range from 250 to 800—from the more than 50,000 to 60,000 Jews who were still alive in Warsaw in April 1943. Of course, the lack of weapons helped to keep those figures low. Still, the majority of the ghetto residents chose to defend themselves from the resumption of deportations by building secret hiding places or bunkers.

We who from the safe vantage point of the future expect heroic gestures from a weakened and terrified people betray an innocence of the fundamental nature of the psychology of despair. The Germans understood it well, which is one reason why, as part of their plan for extermination, they deliberately reduced their victims to a bare minimum of physical and moral energy. We don't need to guess what this must have been like; we have ample testimony from the documents. Lewin himself records meticulously his response when during the mass deportations his wife is sent off to Treblinka. His remorse is genuine; but so is his paralysis. "Eclipse of the sun, universal blackness," he writes:

> My Luba [his wife] was taken away during a blockade. . . . To my anguish there is no prospect of rescuing her. It looks like she was taken directly into the train. Her fate is to be a victim of the Nazi bestiality, along with hundreds of thousands of other Jews. I have no words to describe my desolation. I ought to go after her, to die. But I have no strength to take such a step.[7]

As we survey the spectacle of Lewin remembering his wife, and his inability to do anything to save her, we begin to understand how little an expression like the "redeeming power of memory" can have for him, or for us. Holocaust memory redeems only when it falsifies; Lewin's subsequent reactions to his loss document the efforts of what I call unheroic memory to wrestle with a legacy that has tainted his consciousness with an indelible stain. The following day he writes: "I will never be consoled as long as I live. If she had died a natural death, I would not have been so stricken, so broken. But to fall into the hands of such butchers!"

Lewin speaks of the tragic end of their life of twenty-one years together. But in tragedy, the victim is an agent, or at least a partial agent, in his or her own fate. We know, and as we slowly perceive, Lewin does too, that a main source of his anguish is his failure to find a role for his wife or himself in what happened to her. This fills him with an unappeasable grief: "My soul can find no peace," he records two days later, "for not having gone after her when she was in danger, even though I could also have disappeared."[8] The heroic imagination conjures up all kinds of knightly exploits that Lewin might have attempted, but these are the stuff of Romantic literature or folklore, not life, and though some of us persist in imposing such exploits on the grim reality of the Holocaust, Lewin's humble words remind us how humanly unexceptional most of us are, even in moments of extreme disaster. I think if pressed, Ringelblum might have admitted this too.

"Today is the seventh day since the great calamity that befell me," Lewin writes after a week has passed. "If only I could die and be free of the whole nightmare. But I am still tied to life and it is still difficult for me to take my own life."[9] Whatever we may call his clinging to life, celebrating it as an affirmation of the human spirit, considering the immediate context of his loss, would be to misconstrue his situation and his attitude. Since virtually all surviving victims share a similar kind of loss, it is little wonder that they demur when we ply them with the rhetoric of heroic behavior. They know we do this to shield ourselves, not to praise them.

Lewin's own vista of what lay ahead paid homage to the limitations, not the infinite vitality, of the human spirit. "The burden on our souls and on our thoughts has become so heavy, oppressive," he wrote, "that it is almost unbearable. I am keenly aware that if our nightmare does not end soon, then many of us, the more sensitive and empathetic natures, will break down. I feel that we are standing on the threshold of the intolerable, between existence and annihilation."[10] One can imagine Adam Czerniaków thinking those very words before he swallowed his poison.

These are not options that endear themselves to the contemporary imagination. But if we are to teach this history faithfully, we must heed without flinching the implications of testimony such as Lewin's, written from within the cauldron. Students of the Holocaust need to know what life from the threshold of the intolerable looks like. Lewin did, and he leaves us the legacy of his vision:

> If we ever live to see the end of this cruel war and are able as free people and citizens to look back on the war-years that we have lived through, then we will surely conclude that the most terrible and unholy, the most destructive aspect for our nervous system and our health was to live day and night in an atmosphere of unending fear and terror for our physical survival, in a continual wavering between life and death—a state where every passing minute brought with it the danger that our hearts would literally burst with fear and dread.[11]

If we ask today, sometimes with a faint if self-righteous air of disapproval, why Jews in the camps or ghettos behaved the way they did, the answer, more often than not, lies locked in a heart bursting with fear or dread. It is an answer beyond judgment—but not beyond compassion.

The language of moral evaluation simply does not serve us in situations like the ones I have been describing. Adina Blady Szwajger, the Warsaw ghetto doctor who poisoned her young patients as the Nazis arrived to seize them, survived the Holocaust and recently wrote a personal account of the episode, leaving us with the challenge of interpretation. She was not driven by her own fear or dread, she says, but by the fear and dread of the children, whose plea that she stay with them "until the end" she resolved to heed. She decided that she could best fulfill this pledge by becoming the agent of that end. She thus seized two large containers of morphine (the "poison" that Edelman spoke of), and calmly narrates how she proceeded:

> I took the morphine upstairs. Dr. Margolis [head of the tuberculosis ward] was there and I told her what I wanted to do. So we took a spoon and went to the infants' room. And just as, during those two years of real work in the hospital, I had bent down over the little beds, so now I poured this last medicine down those tiny mouths. Only Dr. Margolis was with me. And downstairs there was screaming because the Szaulis [units of Lithuanian collaborators] and the Germans were already there, taking the sick from the wards to the cattle trucks.
>
> After that we went in to the older children and told them that this medicine was going to make their pain disappear. They believed us and drank the required amount from the glass. And then I told them to undress, get into bed and sleep. So they lay down and after a few minutes—I don't know how many—but the next time I went into that room, they were asleep. And then I don't know what happened after that.[12]

Dr. Szwajger managed to escape from the ghetto and live with false papers on the Aryan side for two years as a courier for the Jewish Fighting Organization, but she confesses that those years "still didn't manage to wipe out any of what had happened the day I gave the children morphine." That's why, she concludes, "I was always different from everybody else. And nobody ever understood this. Everybody thought I'd forgotten about everything and didn't care any more."[13] This is a classic example of what I elsewhere call tainted memory, a concept vital for us to understand if we are ever to assess adequately the legacy with which the Holocaust has smitten our consciousness.

Since the children, like all the other Jews in the Warsaw ghetto, were sentenced to death anyway, does it matter how they died? Knowing what they were spared, we are forced by the circumstances to view the killing of the children as an act of mercy, and this itself reminds us of what the Holocaust has done to the systems of value that we cherished before its advent. The more we immerse ourselves in the personal ordeal of victims such as Abraham Lewin, Marek Edelman, and Adina Szwajger, the more we must adapt ourselves to an idea that in its relentless harshness shares the stage with the painful notion of being sentenced to die—and that is the anguish of being sentenced to live. The postmodern replacement of the death sentence that dominated western thought from Freud to Camus is the life sentence.

What can this mean? Because, unlike other crucial episodes in history such as the French and American revolutions, the Holocaust is an event without a future—that is, nothing better for mankind grew out of it—memory is sentenced to confront it without any relief from expectation. Death is no longer a destiny to be postponed (or transcended), but a constant companion. During their ordeal in the Warsaw ghetto, inhabitants knew that daily survival was merely a respite, not a triumph. And even after the war, people like Marek Edelman and Adina Szwajger realized that little had changed. You don't remake your life after an event like the destruction of the Warsaw ghetto; you simply are sentenced to live with the memory of the ruin.

One of the laudable rhythms of public or private history is that most human beings eventually seem able to resume their lives after a disaster. Superficially, this is true of Marek Edelman and Adina Szwajger too. But if we read beneath the lines of their testimony, we hear a muted theme, and it is here, in what I have called, following Auschwitz survivor and memoirist Charlotte Delbo, the realm of deep memory,[14] that a darker truth emerges: tainted memory leaves a tainted legacy and a tainted life. For Marek Edelman, recalling details of the Warsaw ghetto uprising leads to a bitter discovery: history is not a chronology of events; rather, "historical order turns out to be nothing more than the order of dying."[15]

Dr. Szwajger turned to pediatrics after the war, specializing in tuberculosis in children, pursuing her belief that one becomes a doctor in order to save life. "But somewhere underneath," she admits, "I thought that I had no right to carry out my profession. After all, one does not start one's work as a doctor by leading people not to life but to death." She knows there were reasons for her behavior during that time, but memory cannot appease such knowledge. She is unable to escape the feeling that "along the way something was not as it should have been." And she concludes her formal narrative with a question that continues to echo in our own helpless minds: "Maybe it was too heavy a burden for the rest of my life?"[16]

Possibly we can relieve that burden by adding some of it to our own consciousness. This requires us to accept and generalize Dr. Szwajger's private conclusion: in the history of the Warsaw ghetto, along the way not only something but *everything* was not as it should have been. The Germans were ruthless in their plans for total destruction. The Polish underground's supply of arms to the Jewish uprising leaders was scanty and, as it later turned out, inappropriate: they provided pistols, when the Jews needed rifles and machine guns for the street fighting. Meanwhile, the outside world simply ignored pleas for help. The number of Jewish fighters engaged in the uprising was never more than between 1 and 2 percent of the remaining ghetto population. The majority of those who outlived the search and the subsequent burning of the ghetto were shipped to Sobibor or Treblinka. Few survived. On the fourth day of the uprising, the commander of the Jewish Fighting Organization, the twenty-three-year-old Mordecai Anielewicz, wrote to his Jewish liaison on the "Aryan" side: "I can't begin to describe the conditions under which the Jews are living. Only an elect few will hold out under them. All the others will perish, sooner or later. Our fate is sealed. In the bunkers where our comrades are hiding, it is not even possible to light a candle at night for lack of air. . . ."[17] Two weeks later, the Germans discovered the command bunker; most of its inhabitants, including Anielewicz, died, through either suicide or the poison gas that the Germans pumped into the bunker after blocking the exits. A handful, including Marek Edelman, managed to make their way out of the ghetto through the sewers to the "Aryan" side. The ghetto itself was doomed.

What was the Warsaw ghetto's role in the history of the Holocaust? We will go on interpreting it for decades and perhaps generations, with the help of the immense number of documents from Jewish, Polish, and German sources. Simply celebrating the exploits of its courageous, if futile, defenders seems to me a hopeless effort at self-delusion. I think Claude Lanzmann understood this as well as anyone, since he chose to end his nine-and-a-half-hour cinematic epic on the destruction of European Jewry, *Shoah*, with the voices of two figures who were central to the ghetto's defense. What they say allows Lanzmann to deflate the desires of his viewers, many of whom, when the subject of the Warsaw ghetto was finally raised in the film, must have expected some heroic relief from their long and dismal encounter with unnatural death. But they were to be disappointed. Lanzmann chose to let the ruins of memory prevail. One of his voices, Simha Rotem, known as Kazik, prefers facts to heroism:

> I don't think the human tongue can describe the horror we went through in the ghetto. In the streets, if you can call them that, for nothing was left of the streets, we had to step over heaps of corpses. There was no room to get around them. Besides fighting the Germans, we fought hunger, and thirst. We had no contact with the outside world; we were completely isolated, cut off from the world. We were in such a state that we could no longer understand the very meaning of why we went on fighting.[18]

That search for meaning was complicated by Rotem's description of the situation outside the ghetto: "In Aryan Warsaw, life went on as naturally and normally as before. The cafés operated normally, the restaurants, buses, streetcars, and movies were open. The ghetto was an isolated island amid normal life."[19]

And there it will remain throughout history, unless we allow it to penetrate our consciousness and shatter the rhetorical shield of heroism that protects us. The other voice with which Lanzmann ends his film is that of Itzhak Zuckerman, second-in-command of the Jewish Fighting Organization in the Warsaw ghetto and, along with Edelman, the only surviving member of the leadership. Asked by Lanzmann to comment on his memory of the event, Zuckerman succinctly replies: "I began drinking after the war. It was very difficult. . . . you asked for my impression. If you could lick my heart, it would poison you."[20]

If normal memory is an internal ordering of images from the past, then tainted memory is an internal disordering of those images, and Zuckerman's uncommon response to his Warsaw ghetto experience reveals the origins of that taint. It is an austere and vexing legacy, but the Holocaust, when truly faced, offers us little else.

1993

4

Ghetto Chronicles: Life at the Brink

lthough diaries of the deathcamp experience were still being dug up in the ruins of the gas chambers at Auschwitz as late as 1962, they were few in number, often fragmentary, and usually in poor condition. The most detailed records of Jewish existence during the Holocaust are to be found in archives and chronicles from ghettos like Warsaw, Lodz, and Kovno. Carefully hidden or buried at the time, they surfaced after the war to furnish vivid accounts of how an oppressed and deceived people managed the daily rhythms of life as the Germans contrived their relentless migration toward death.

Reading these narratives is a chastening task. More than any other kind of Holocaust writing, they force us to forgo the wisdom of hindsight and to tackle inquiries more modest than ones urging us to ask why the victims did not foresee their fate and prepare to resist it. Retrospective vision seeks to integrate reality, even though all the evidence suggests that most ghetto residents faced a constant *dis*integration, through hunger, illness, or despair, of the family and community supports that ordinarily help us to thrive. Looking backward, as Tolstoy insisted in *War and Peace,* always makes history seem less random than it actually was. If we enter the world of the ghetto through the eyes of its witnesses, many of whom left their written testimonies as their sole legacy to the future, we can begin to imagine how a constant wavering between expectation and futility drained the self of the inner strength that sustains one in more normal times.

For example, among the papers rescued from the Lodz ghetto was the diary of an anonymous young girl who in regular entries measured the quantity of dekagrams (similar to our ounces) of food that she consumed each day. Of course, she had other themes, but this was a major focus of her journal, exposing an inner gnawing that invaded everything else she wrote about:

> *Saturday, 7 March 1942*
> Beautiful sunny day today. When the sun shines, my mood is lighter. How sad life is. One yearns for a different life, better than this grey and sad one in the ghetto. When we look at the fence separating us from the rest of the world, our souls, like birds in a cage, yearn to be free. . . . How I envy the birds that fly to freedom. Longing breaks my heart, visions of the past come to me. Will I ever live in better times? After the war will I be with my parents and friends? Will I live to eat bread and rye flour until I'm full? Meanwhile, hunger is terrible. Again we have nothing to cook. I bought 1/4 kg. [about 1/2 pound] of rye flour for 11 1/2 rm [reichsmarks, the currency at the time]. Everybody wants to live.[1]

The nostalgia that invades so many of our memories of the Holocaust from our own safe perspective prompts us to imagine that despite severe ghetto conditions, at least families clung together, serving one another's needs when help from outside failed. Unfortunately, much of the evidence draws us in less happy directions. The same young diarist writes a few days later:

> Today I had an argument with my father, I insulted and even cursed him. And this was because yesterday I weighed 20 dkg of noodles but this morning took a spoonful for myself. When father came back at night, he weighed the noodles again. Of course there was less. He started yelling at me. He was right, of course; I had no right to take for myself the few precious dekagrams of noodles Mr. Chairman [that is, Chaim Rumkowski, Elder of the Jews in the Lodz ghetto] gives us. I was upset and I cursed him. Father just stood at the window and cried like a child. No stranger ever abused him like I did. Everybody was at home. I went to bed quickly without touching supper. I thought I would die of hunger.[2]

Familiar vocabulary like "adolescent immaturity," "selfishness," or "lack of compassion" proves futile to explain such behavior. These responses now appear as luxuries of conduct resulting from a full belly. What the diaries and journals of ghetto life, written at the time of the ordeal, help us to do is to enter into the daily experience of deprivation, atrocity, and constant fear that were sometimes literally the only nourishment that the victims had. The danger we must guard against is the impulse to blame *them* for the way they responded.

One of the many values of these documents is that they give us a far more vivid sense of the *source* of this misery than do retrospective accounts, which are less exact in recapturing the minutiae of feelings as they occurred. We are so transfixed by the crime, we tend to lose touch with the criminals. The following excerpt, from the same anonymous diarist, brings us back to the cause of such wretchedness, identifying the agents. In one of the marketplaces, she writes,

> German workers were repairing the electric wire and were pitching a tent. A woman passed by the tent. One of the workers pushed her to the ground and started beating and kicking her. People ran, scared, in every direction. Nobody said a word. For each word not to their liking, hundreds of Jews could perish. How tragic is our life, how humiliating. We are treated worse than pigs. We Jews of the ghetto, we work so hard, we help them in the war, making beautiful things from rags—military uniforms, rugs, everything a person needs. They treat us worse than slaves. And this is life. Isn't death better?[3]

This somber question is more than rhetorical. It sheds light on one of the issues we in our innocence continue to explore: Why didn't the victims do more to keep themselves alive? One answer we screen ourselves from hearing is that occasionally, because of the unbearable persecutions they were subject to, they preferred not to.

But this is not the usual case. Although the threat and then the prospect of death began to infiltrate their consciousness like a rabid virus, most victims clung to life with desperate tenacity. Another young diarist, who perished of tuberculosis in the ghetto at age nineteen, wrote: "Yesterday a student in our class died from hunger exhaustion. Because he looked awful, he was allowed to have as much soup in school as he wanted, but it didn't help. He was the third victim of starvation in the class."[4] Such details give us a glimpse of the ceaseless if unequal struggle between everyone's wanting to live and the firm resolve of a regime determined to see them die.

But the ghetto chronicles also address much more comprehensive matters, even though sometimes indirectly. We now know what the Jews in the ghettos did not: that the Germans planned to concentrate them there only as a stage in their eventual extermination. We learn much about the Jewish administration of the ghettos from these documents, and through them some details about the German officials pulling the strings from the wings. If Chaim Rumkowski turns out to be a pathetic megalomaniac, deceiving and self-deceived, and Adam Czerniaków, head of the Jewish Council in the Warsaw ghetto, a helpless man of integrity, less deceiving than deceived, reading between the lines from the vantage point of today enables us to see that in spite of their efforts to preserve life, both were little more than pawns in the hands of the Germans, who maneuvered them with lies until they were ready to strike.

These documents, in other words, give us insight into the minds and wills of men and women who want to live but have been deprived of the means. Once we realize that *nothing* Rumkowski or

Czerniaków might have done could have saved the lives of the Jews of Lodz or Warsaw, we can perhaps approach judgment of their behavior with a more lenient stance. Rumkowski may have believed to the end that he had sacrificed the "few" to save the many; in this, he has his defenders and his foes. In Czerniaków's opinion, Rumkowski was "replete with self-praise, a conceited and witless man."[5] Refusing to preside over the deportation of Warsaw's Jews to Treblinka, Czerniaków himself took poison and ended his life. This gesture, too, has earned rebuke and admiration. The more we immerse ourselves in the daily ordeal of the ghetto residents, leaders and ordinary inhabitants alike, the more we see that they were all faced with a choice between *im*possibilities—no meaningful choice at all. Even the courageous Warsaw ghetto uprising was a venture in behalf of an impossibility: it could save almost no one. Its fighters were not eager to die; they too preferred to live. The Germans would not allow it.

If the picture emerging from these ghetto documents resembles a moral labyrinth more than the serene spiritual vista we often hear about, this results from something inherent not in the victims' natures, but in the malicious intentions of their killers. Although some of the German officials in charge of the ghettos (prodded by the Labor Ministry) agreed to keep alive as long as possible those Jews producing uniforms and munitions for the military, they also cooperated with the SS as they systematically reduced the size of the ghettos by deporting less "useful" members of the community to the deathcamps. Offering false assurances to the beleaguered leaders, who passed them on halfheartedly to their desperate people, the oppressors preyed on their victims' instinct for hope. Rumors of impending improvement in these pages expel darker hints of doom, as the Jews cling to the shreds of life even as the fate of friends and family members gnaw at their will to survive.

Only extreme psychological naïveté might lead us to be surprised. Among the most hated groups in certain ghettos, after the SS, were the Jewish police, appointed by the administration and used to maintain order, but also to round up candidates for deportation. Although many Jews refused to serve in such units, more did not, and for obvious reasons: they and their families were temporarily exempt, higher up in the hierarchy of those who felt immune from persecution. In the end, predictably, they too (along with council members and Elders like Rumkowski himself) were shipped to their death. The Germans never planned to spare anyone. Why didn't the Jewish police decline to do the dirty work of the Germans and volunteer to go first themselves? This too is a naive question, born of our privileged sense of being well fed and unthreatened. One impact of these writings is to carry us back into the reality of that milieu, governed by daily fear and hunger, when most human beings were stripped of the luxury of caring for anyone but one's own.

Perhaps the bitterest truth we encounter in these ghetto records is how easily the concept of "one's own" shrank into the narrow identity of one's *self.* This does not mean that members of the Jewish Council in Kovno, Czerniaków in Warsaw, and even Rumkowski in Lodz did not argue or negotiate with their German masters to better conditions for their fellow Jews, often at considerable risk to themselves. They reasoned and pleaded to increase rations, reduce the number of deportees, improve housing and sanitation—and sometimes they succeeded. But the Germans' appeasement had nothing to do with pity; it was all part of a covert deception to mislead their victims and abate suspicion, drawing on the pliant human impulse to be beguiled, when the issue is one's imminent death.

When 27,000 Jews from the Kovno ghetto are summoned on October 28, 1941, to a massive "roll call" from which 10,000 would be selected for execution, we gain an unforgettable view of the miserable cruelty of survival:

> The square was surrounded by machine-gun emplacements. Rauca [a Gestapo official] positioned himself on top of a little mound from which he could watch the great crowd that waited in the square in tense and anxious anticipation. His glance ranged briefly over the column of the Council members and the Jewish Ghetto police, and by a movement of his hand he motioned them to the left, which, as it became clear later, was the "good" side. Then he signaled with the baton he held in his hand and ordered the remaining columns: "Forward!" The selection had begun.
>
> The columns of employees of the Ghetto institutions and their families passed before Rauca, followed by other columns, one after another. The Gestapo man fixed his gaze on each pair of eyes and with a flick of the finger of his right hand passed sentence on individuals, families, or even whole groups. Elderly and sick persons, families with children, single women, and persons whose physique did not impress him in terms of labor power, were directed to the right. There, they immediately fell into the hands of the German policemen and the Lithuanian partisans, who showered them with shouts and blows and pushed them toward an opening especially made in the fence, where two Germans counted them and then reassembled them in a different place.[6]

The Germans were interested in totals, not persons, so that if the figure were correct, the identity of the victims would be trivial. In the chaos of the moment, much shifting from the "bad" side to the "good" side occurred; in addition, members of the Jewish Council were able to convince the SS to free some of those who were selected if they were vital to the order of the community. But today who can avert his face from the ruthless German algebra dictating that every life saved must be paid for with the death of someone else? And who is morally responsible for the feat of balancing that infernal equation?

After this mass selection of 10,000 doomed men, women, and children, who were marched to the nearby so-called Ninth Fort at Kovno and shot, the "reprieved" remnant returned to the ghetto. We know what they survived *from;* what had they survived *to?* Memory of their loss intervened to forestall relief at their escape, a rhythm prototypical of the Holocaust experience. Immediately, crowds thronged to the quarters of Dr. Elchanan Elkes, chairman of the Kovno Jewish Council, pleading with him to save those family members who had been taken away. He scrawled down as many names as he could, made his way to SS headquarters, and miraculously managed to gain permission to rescue 100 from those on their way to their death. How was he to choose the 1 percent to be spared, while the other 99 percent perished? Armed with his approved petition, filled with despair and hope, he reached the ranks of the condemned, but his presence caused such turmoil among them that the Lithuanian guards, fearful of a rebellion, beat him to the ground with their rifle butts, and he was carried bleeding back into the ghetto.

Human beings should not have to ask themselves to make such choices—what I call choiceless choices, because whatever you choose, somebody loses—shorn of dignity and any of the spiritual renown we normally associate with moral effort. Nazi rule in the ghettos was designed to exploit and humiliate the ethical bias of their victims, even poisoning the value system implicit in language. If Elkes had succeeded in his mission, could he possibly have been considered "heroic"? Or would those he might have "saved" merely abide as hostages to the SS desire to nurture illusion among the living remainder? We know that the SS's willingness to release 100 victims had nothing to do with respect for the petitioner, who was trapped by a form of malice he was unable or unwilling to envision. What appeared to be courageous resistance from the Jewish point of view (and all of this, of course, is to Elkes's credit) is nothing but a contemptuous temporary forbearance within the Nazi scheme of mass murder. Both choice and power were in the hands of the murderers, but failure to grant this caused some Jewish leaders to cling to a belief in the meaning of options whose significance we are still trying to appraise.

The most notorious example of this dilemma is the "Give Me Your Children" speech delivered by Elder of the Jews Chaim Rumkowski in the Lodz ghetto on September 4, 1942. Evidently convinced, as noted earlier, that by surrendering the few he could preserve the many, Rumkowski conceded, "I must cut off limbs in order to save the body itself." But those "limbs" were all the children in the ghetto under the age of ten (as well as the ill and the elderly over sixty-five). Did Rumkowski worry that without its limbs, the body would lose much of its vital force? If the fear of death were so overwhelming then, what would the fear of *life* be like for those who survived the war, once they had time to reflect on the loss (that is, the children, the ill, and the elderly) that had ensured *their* continuing existence? The questions that come so easily to us today may have seemed less urgent to a community robbed of its future, mired in the terror of the single day. Only a strenuous leap of the imagination, aided by the words of the chroniclers themselves, can lift us out of the flow of time and fix us in the arrested moment when Rumkowski's audience stood frozen in disbelief on that September afternoon.

"A grievous blow has struck the ghetto," Rumkowski began. "They are asking us to give up the best we possess—the children and the elderly." By seeking their cooperation, did Rumkowski want to spare his people in Lodz the horror and brutality of the Warsaw ghetto, where even as he spoke the Germans were nearing the end of their vicious roundup and deportation of more than 300,000 Jews to Treblinka? Was his rhetoric his own, or was he duped by Nazi promises? "So which is better?" he asks. "What do you want: that eighty to ninety thousand Jews remain, or, God forbid, that the whole population be annihilated?" This language is unmistakable; Rumkowski must have known the impending fate of the children, and the ill and elderly who were to join them (the Germans had demanded 24,000, but Rumkowski announced, and apparently believed, that he had bargained the figure down to "only" 20,000). Encouraged by his Nazi masters to trust in options, Rumkowski found his reasoning irreproachable: "[P]ut yourself in my place, think logically, and you'll reach the conclusion that I cannot proceed any other way. The part that can be saved is much larger than the part that must be given away."[7]

Today, we are baffled by Rumkowski's use of language. How can a mother and father be expected to decide that it is "logical" to save themselves by surrendering the lives of their children? On the surface, Rumkowski's appeal seems to display genuine anguish:

> I have no thought of consoling you today. Nor do I wish to calm you. I must lay bare your full anguish and pain. I come to you like a bandit, to take from you what you treasure most in your hearts! I have tried, using every possible means, to get the order revoked. I

> tried—when that proved to be impossible—to soften the order. Just yesterday, I ordered a list of children aged nine—I wanted, at least, to save this one age group, the nine- to ten-year-olds. But I was not granted this concession.[8]

How does one negotiate for the nine-year-olds, knowing that those eight years old and younger will be sent to their death? The mystery we will never solve is how the Germans managed to persuade Rumkowski—and through him, many of the Jews he addressed—to delude himself into believing that he was acting within a framework of *choice,* thus preserving the faculty that endowed both him and his audience with a measure of freedom. In this atmosphere of mass destruction, Rumkowski uses words like "common sense" as if they still had meaning:

> There are, in the ghetto, many patients who can expect to live only a few days more, maybe a few weeks. I don't know if the idea is diabolical or not, but I must say it: "Give me the sick. In their place, we can save the healthy." I know how dear the sick are to any family, and particularly to Jews. However, when cruel demands are made, one has to weigh and measure: who shall, can and may be saved? And common sense dictates that the saved must be those who can be saved and those who have a chance of being rescued, not those who cannot be saved in any case.

By nurturing this delusion of "rescue" to the end, the Germans gained Rumkowski's cooperation. They were able to do this because he refused or was unable to believe in two things: his own helplessness, and the unrestrained barbarity of the Germans. He could not admit to himself that they planned to kill every Jew in the ghetto—all, without exception. That made no "sense" to him, any more than it does to us today, when we continue to resist the notion that the murder of European Jewry was totally without sense or meaning. His ego needed to feel that he was still in control of some portion of his surrounding reality: "Give into my hands the victims, so that we can avoid having further victims, and a population of a hundred thousand Jews can be preserved. So they promised me: if we deliver our victims by ourselves, there will be peace."[9] *So they promised me:* these words, together with Rumkowski's pathetic appeal to "reason and conscience," confirm how much the Holocaust represented the sabotage of language as well as the destruction of lives. Reason and conscience were as distant from that event as Saturn and Jupiter are from the earth. I would like to say that we now realize how alien is the world where innocent children are murdered from the one where reason and conscience dwell—but I'm not so sure that we do. What must Rumkowski have felt as he and the Jews he presumed to save on that fatal September day entered the gas chambers of Auschwitz themselves two years later? We can try to imagine, though we will never know.

If we shrink now from Rumkowski's request for the ill and the children, and from their parents' eventual consent to surrender them, we display a habit of mind that flourishes in a climate free of the terror that prevailed then. Some survivors of the Lodz ghetto insist that they would be dead today if Rumkowski had not delayed the date of their deportation through such "arrangements." But Czerniaków in Warsaw committed suicide rather than ask his fellow Jews to give up *their* children for "resettlement." He was left without illusions and without hope, and that eroded his will to live. He apparently saw what Rumkowski balked at accepting: the part that was saved was *the same as* the part that was given away, all being candidates for destruction, some now, some later. This helps us to understand the bitter words, so relevant here, of the narrator in one of Tadeusz Borowski's Auschwitz stories, writing to his fiancée in another part of the camp: "Never before in the history of mankind has hope been stronger than man, but never also has it done so much harm as it has in this war, in this concentration camp. We were never taught how to give up hope, and this is why today we perish in gas chambers."[10] Rumkowski's error was in assuming that sound premises would ensure stable outcomes. He refused to concede—and he was not alone in this—that the Third Reich buried premises as well as people.

They buried people twice, once before their death, and once after, and this is perhaps the most vicious of their many crimes. How is it possible to bury a man while he is still alive? How is it possible to make innocent Jews feel that they are murderers too? The Germans managed to find a way. We try to imagine what it must have been like for parents who responded to Rumkowski's plea and in order to save the "many" gave up the "few" for deportation, their own children under the age of ten. Those of us who are parents or grandparents ourselves can't imagine it—and then we are helped by the anguished words of a father who was there, words preserved in a Yiddish fragment written on the back of four pages of a ghetto soup-kitchen record:

> *Wednesday, September 8, 1942*
> Yesterday I lost Mookha, my sweet little daughter. I lost her through my own fault, cowardice, stupidity and passivity. I gave her up, defenseless. I deserted her, I left the 5-year-old child, did not save her, and I could have done it so easily. I killed her myself because I didn't have the least bit of courage. . . . I, her father, did not protect her, I deserted her because I feared for my own life—I killed. . . . I am

> broken, I feel guilty, I am a murderer and I must atone, because I won't find peace. I killed my child with my own hands, I killed Mookha, I am a killer, because how can it be that a father deserts his own child and runs away? How can he run away and not save his own child?[11]

He becomes a spokesman for what I elsewhere call the humiliated self, filled with remorse for a deed of which he is entirely innocent. Had he survived—and he did not, blessedly, I am inclined to say—how would he have entered into his future with rekindled hope? Surely we can say that such a man has died twice.

The events described in these journals and diaries oblige us to forfeit the cherished idea that no matter how cruel the circumstances, civilized men and women always achieve dignity by making moral choices and abiding by the results, beneficial or not. When the challenge to being is the minimal demand of remaining among the living, the test of character has little to do with spiritual stature. These texts cause us to revise our notion of what reasonable creatures are capable of under duress. The anonymous young diarist quoted earlier records her horror when one day she surprises her father in the act of furtively eating part of her mother's bread ration. She is dismayed, but less quick to condemn, since she knows how ceaseless hunger has infected her own will to live. The clash between family and community duties and the needs of the self was constant, remorseless, and ultimately beyond reconciling. The appetite of the German death machine for more victims from the ghettos was unappeasable. The "lucky" ones died before deportation; we know the doom of the others. Choice had little to do with their fate, which took them beyond the frontiers of moral endeavor.

If this is a harsh truth for us to absorb, how much more painful must it have been for the victims, who had to grant in their final misery that there were no limits to the evil their killers were capable of, while the good that might aid them was distressingly finite, drained by an ever-dwindling number of fragile chances to stay alive.

1992

5

Cultural Resistance to Genocide

he expression "cultural resistance to genocide" consoles us even before we attempt to understand its implications. "Genocide," a term coined in 1944 to describe the physical destruction of European Jewry by the Nazis in World War II, threatens to engulf us with the finality of its achievement: nearly 6 million Jewish victims, with more in the offing, had not Hitler's defeat ended the carnage. "Resistance to genocide," as both concept and fact, restores a measure of dignity to the victim (and bestows crucial psychological satisfaction on us as retrospective spectators of the event). There is no need to rehearse here the courageous episodes in Warsaw and other ghettos, or the uprisings in Auschwitz, Sobibor, and Treblinka. Ultimately, they did not stop the slaughter; but they saved some lives, and they left us with gratifying proof that when the situation was hopeless, men and women could stand up (the basic meaning of "resist") against impossible odds and try to escape or choose death rather than submit quietly to their doom. Physical resistance to genocide may have been infrequent during the Holocaust (chiefly because the Nazis combined deception with brutality to discourage it), but the *meaning* of the gesture is not difficult to understand since examples of opposition to the genocidal impulse were not unique to the Holocaust experience.

Adding "cultural" to the sequence of words, however, introduces complications and eventual contradictions that we may not so readily resolve. It is easier and, I hope to show, more accurate to speak of cultural *witness* to genocide (or atrocity); for this we have a long and respectable tradition of antecedents. Goya's *Disasters of War,* Otto Dix's macabre visions of World War I, Picasso's acclaimed *Guernica*—all used a form of culture, artistic expression, to testify to inhuman horrors that might otherwise remain inaccessible to the human imagination. But we have little evidence that Goya's haunting drawings modified the nationalistic enthusiasms of future war-committed generations. We do know that the Hitler regime silenced the pacifist impulse of Otto Dix, who retired to the country and painted landscapes for the duration of the war. And no one has ever claimed that Picasso's painterly wrath suppressed a single atrocity in Franco's Spain or Nazi Germany. Nevertheless, their art, like the poems and songs from the partisans and the ghettos, the writings and paintings and drawings from Theresienstadt and other camps, forms an indispensable contribution to our understanding of the consequences of atrocity. This art resonates beyond the verbal chronicles of survivors and historians; it afflicts our desire to redesign hope from the shards of despair with the vision of an anguish that is recordable but not redeemable by the energies of cultural effort.

How much of the origin of a phrase like "cultural resistance to genocide" springs from what I have called "our desire to redesign hope from the shards of despair"? The ordeal of European Jewry is over for the victims; it will never be over for us. The pendulum of response to this ordeal oscillates between stories of heroism and accounts of human agony, but the duration of the swing in either direction depends on the needs of the spectator as much as on the details of the event. The Holocaust evokes various reactions, ranging from confronting the worst to coping with the manageable, with frequent alternations between the two. What we call "cultural resistance" to mitigate our horror at the event may have been for the victims nothing more than a similar alternation between confrontation and coping: sketches of corpses and drawings of deportations, on the one hand, and, on the other, portraits and landscapes of normalcy, preserving the inherent dignity in the faces of prospective victims soon to meet a harrowing if unexpected doom. Using their tools and talents, such artists etched on the tablets of history a permanent record of their experience. But how did it injure the enemy, scornful of the very idea of Jewish culture? And how did it save Jewish lives?

The intent of culture, we may assume, is to preserve and enrich intellectual, moral, and spiritual existence. The purpose of genocide is to destroy physical life. The weapons of each are entirely different, and certainly unequal. Josef Bor in *Terezín Requiem* reconstructs the story of a special performance of Verdi's *Requiem* prepared by order of the commandant of Theresienstadt for camp personnel and some visitors from Berlin, among whom was Adolf Eichmann himself. The conductor was Rafael Schächter, a prominent Czech deportee, who drew his musicians, chorus, and soloists from other talented Jewish artists among the prisoners. Rehearsals were a temporary sanctuary for their insulted bodies and souls, a brief, officially sanctioned respite of normalcy from the pervasive chaos of camp routine. The visitors arrived; the performance went off as scheduled, somewhat diminished by the insistence that it last no longer than an hour; the Nazi guests departed. But the irony of the choice, Verdi's *Requiem*, seems to have been lost on no one, since two days later Schächter and his entire group were transported to Auschwitz, where virtually all of them perished in the gas chambers.

Verdi's *Requiem* is a classic example of what we call culture; and the fate of the performers, a reflection of what we call genocide. The problem arises with the connecting word "resistance," and I would like to suggest the possibility that its importance may have been exaggerated by *us*, the heirs and descendants of the Holocaust experience, instead of its having been recruited as an accurate description of their activities by the victims and survivors of the event. Early during his rehearsals, before a special performance had been arranged, according to Bor's version, Schächter excitedly informed his artists that camp officials had promised not to deport the members of this group. General rejoicing prevailed at this unexpected gift of salvation. And for the time being, the cynical Germans kept their promise; but a few days later, they deported the performers' parents, spouses, and children! Most of them chose to join their families on the journey to Auschwitz, and the "resistance" value of the Verdi enterprise was suddenly deflated by that unpredictable turn of events. Schächter worked furiously to recruit replacements, and finally succeeded.[1] It should be clear to us, however, how precarious is the dignity one achieves by affirming the value of art as a form of resistance under such circumstances. Lest we falsify the ordeal of the victims and survivors in order to appease our own (real) pain when confronting their anguish, it seems crucial to appreciate the limitations of an expression like "cultural resistance," and not attribute to it a power it did not possess.

When art records what Leo Haas, one of the surviving painters of Theresienstadt, has called "an existence unworthy of human beings" (viciously transformed by Nazi racial ideology into *das lebensunwerte Leben*, a life unworthy of being lived), how are we to identify it with cultural (or spiritual) resistance, as it is so often called? The challenge, though not futile, is certainly strenuous, since we are asked to connect the humiliation of man, which is the subject of so many of these paintings, with an attitude that somehow transcends that humiliation. We look at sketches of starving Jews, of crowds awaiting deportation to the deathcamps, of desolate children, of executions, and suddenly "resistance" sounds irrelevant and "culture" itself a term from an antiquated vocabulary, implying harmony, order, disciplined thought, beauty, and form, thumbed from a thesaurus that never heard of the world of Theresienstadt or the events that transpired there. The process of perception that gradually liberates in us a creative response to the created work, a process that unfolds when we contemplate most art, does not operate here. Where in our own imaginative life are we to record and absorb such documentary materials, as Leo Haas liked to call the efforts of concentration camp artists? Perhaps future generations, more accustomed to atrocity than we, will find it easier; for them, the shock of recognition may not be so disabling. But if a time arrives when art can celebrate such "truth" with ease, and audiences can appreciate it with a deferential nod to a new category called "cultural resistance," maybe we should be grateful today to be able to wrestle with our unarticulated pain.

Alfred Kantor, who made hundreds of sketches and drawings in Theresienstadt, Schwarzheide, and even Auschwitz (according to his own testimony), and then destroyed most of them for fear of being discovered, reconstructed them from memory in the weeks following his liberation. He recognized the fragility and the danger of what we call cultural resistance, and refused to heroize the activity. Twenty-five years after the war, he published his work as *The Book of Alfred Kantor.* It would be tempting to argue that Kantor's labors as an artist during his camp experience enabled him to retain his sense of dignity and kept him alive. But Kantor makes no such claim. Looking back, he speaks of the *psychological* value of his secret artistic endeavors:

> My commitment to drawing came out of a deep instinct of self-preservation and undoubtedly helped me to deny the unimaginable horrors of that time. By taking the role of an "observer" I could at least for a few moments detach myself from what was going on in Auschwitz and was therefore better able to hold together the threads of sanity.[2]

Far from praising culture as a way of *resisting* genocide, Kantor speaks of his art as a form of evasion, a technique for coping through denial of his own possible doom, rather than through confrontation.

But Kantor is unequivocal about his personal capacity to endure. Sketching may have nourished his psychological needs, but it did not feed his wasting body—or anyone else's. Despite his drawings, 800 of the 1000 men transported with him from Auschwitz to Schwarzheide were dead by the day of liberation. Kantor had the uncanny good fortune to have a sister in Prague who was untouched during the war because she was married to a non-Jew. She sent him packages in Auschwitz, and unaccountably some of them were passed through by the SS. He was able to smuggle out a letter to her from Schwarzheide, and soon he was receiving weekly packages there. "Even more than Auschwitz," he says, "these packages were crucial to my survival.

Without this extra nourishment, I could not have endured the months of hard labor at the factory."[3]

If Kantor does not celebrate the survival or resistance value of his drawings and watercolors, why should we? He was shrewd enough to destroy them when he realized that they endangered his life. The painters of Theresienstadt were less careful. When the Nazis discovered their hidden work, they did not marvel at the quality of achievement; they rounded up the artists, beat and tortured them, then deported them to Auschwitz. Unfortunately, genocidal practice fails to distinguish between cultural impulse and Jewish disobedience. The art that survives appeals to us to see the inhuman behind the facade of the human; despite courageous moments of Jewish dignity, the victims lacked the power to oppose the German determination to erase them from the face of the earth. It is an art of commemoration, asking us not to acknowledge heroic lives, but to mourn melancholy deaths.

And even "deaths" is not an exact term for the doom of Hitler's victims, who lacked our retrospective vision of what awaited them at the end of their boxcar journeys. Auschwitz survivor Jean Améry's famous formulation that "no bridge led from death in Auschwitz to *Death in Venice*,"[4] provides a seminal insight into the *dis*continuity between culture and genocide during the years of the Third Reich. What had the high artistry of Thomas Mann's obsession to do with the bond between culture and death, or Gustav Aschenbach's "normal" dying to do with the vocabulary of atrocity in the deathcamps: "selection," "extermination," "cremation"? Culture left one *un*prepared for such a doom, and Améry in a single instant repudiated the value of years of study of German romantic literature. Mann's pre–World War I novella, with its vision of the primal destructive energies concealed by art, may have echoed ancestral voices prophesying war, but it did not anticipate gas chambers and the genocide of European Jewry. Améry had to search elsewhere for armor to protect his now naked sensibilities. The culture of the past mocked the moral anarchy of the present.

But all potential victims did not share Améry's hard-won recognition, and it is not difficult to understand why. Cultural traditions furnish a certain security and even sanctity to a life otherwise sundered from the normal props of existence. Few were able to endure on a diet of mere blank terror. The illusion that a familiar past could prepare one for an unknown future never lost its magic appeal to imaginations that had nothing else to rely on. If we continue to respect such attitudes among the victims, we do so perhaps because we can discover no reliable alternatives ourselves. There is nothing spiritually attractive about a vacuum. Hence Janusz Korczak's legendary decision to have the children of his orphanage in the Warsaw ghetto perform a play a few weeks before their expected deportation to Treblinka continues to excite admiration, as a stark example of cultural resistance to genocide, though a less appropriate choice than Rabindranath Tagore's *The Post Office* could hardly have been found. Nevertheless, this sentimental drama about a dying Hindu boy awaiting a letter from his "King," written decades before World War II, produced a "staggering impression" on its audience, according to the editor of Korczak's *Diary* (though, of course, they could not have anticipated the exact form of atrocity awaiting the children in Treblinka).

When after the performance Korczak was asked why he had chosen this particular play, he is reported to have answered that "finally it is necessary to learn to accept serenely the angel of death."[5] The boy in the play was indeed carried off by the "angel of death," a traditional notion of private dying that has long consoled men in their bereavement. Mann's Aschenbach expires with comparable tranquility, while sitting on the beach at Venice. But not many of the victims of Hitler's deathcamps, to say nothing of their survivors, would agree with Korczak's description of their approaching doom as a summoning by the angel of death. The "demon of destruction" might be more apt. Korczak, innocently enough, retreated behind the barricades of an ancient vocabulary; if this was cultural resistance, it was ineffectual against the formidable enemy who lay in wait for him.

Whether it was effectual in preparing Korczak's orphans for their doom, no one can say for sure; but there is much evidence that it was not. Certainly the insane and terrifying *Totentanz* that greeted the children when they disembarked at Treblinka bore no resemblance to the quiet death scene of the Hindu boy in Tagore's *The Post Office*. We need not guess here: a survivor of Treblinka, Yankel Wiernik, has vividly described what happened to children in that deathcamp:

> All through that winter, small children, stark naked and barefooted, had to stand out in the open for hours on end, awaiting their turn in the increasingly busy gas chambers. The soles of their feet froze and stuck to the icy ground. They stood and cried; some of them froze to death. In the meantime, Germans and Ukrainians walked up and down the ranks, beating and kicking the victims.
>
> One of the Germans, a man named Sepp, was a vile and savage beast, who took special delight in torturing children. When he pushed women around they begged him to stop because they had children with them, he would frequently snatch a child from the woman's

> arms and either tear the child in half or grab it by the legs, smash its head against a wall and throw the body away. Such incidents were by no means isolated. Tragic scenes of this kind occurred all the time.[6]

What could Korczak's orphans have thought or felt when they discovered at the final moment of their lives that their stirring confrontation with the angel of death in the play in the ghetto had not even hinted at the looming demon of extermination in the deathcamp? We will never know, and it is irreverent to speculate. But we *do* know that the consolation of private dying, supported by friends and family (as in Tagore's drama), is so contrary to the actuality of genocide that the two experiences seem to belong to entirely different realms of being. To regard Korczak's use of the Tagore play to prepare his charges to meet death serenely as an expression of cultural resistance to genocide, given the subsequent events, is to risk using a heart-warming phrase to deflect our attention from the merciless and chilling murder of these victims.

Just as Rafael Schächter's performance of Verdi's *Requiem* saved no lives, so the production of *The Post Office* in the Warsaw ghetto had no effect whatsoever on the physical fate of Korczak and his orphans. Yankel Wiernik's testimony suggests that it could not have furnished much insulation against terror after their arrival in Treblinka. Nevertheless, the impulse to cultural expression from the midst of irretrievable loss was never silenced. Schächter, Korczak, writers like Yitzhak Katzenelson and Abraham Sutzkever, composers and painters in Theresienstadt who produced hundreds of sketches and paintings, some of great distinction, continued to shore up the ruins of their collapsing world with creative activity and works of the imagination. We may label their legacy a form of resistance, but this is a value-laden word, implying the possibility of an immediately beneficial consequence, a gesture of affirmation or defiance that might alter the condition of the poet, the writer, the painter. But they were the first to realize how self-delusive was such an expectation. Each work, each cultural effort, reflects not defiance but a basic human need to interpret the meaning of one's experience, or to pierce the obscurities that shroud it in apparent meaninglessness. They offer us a glimpse into an arena of terror we can never enter ourselves, a distress we can never share, an inexpressible grief, an incommunicable doom. They challenge us with the difficult awareness of how human beings, hopefully or hopelessly, reacted to the threat of the inhuman.

Some examples may illustrate that challenge. One of the paintings left behind by Leo Haas represents a crowd of gaunt men and women in an attic in a house in Theresienstadt. They are leaning against walls, lying down, dozing, sleeping, or simply staring. They are weary and forlorn, but most of all, they seem to be waiting. We do not know exactly for what, but the drawing is called *Expecting the Worst,* and we deduce that they are soon to be deported to the east. Although they are a crowd, no one seems to be in contact with anyone else. The reputed mutual support that helped victims endure their grief is nowhere in evidence. Each human being is alone with his or her fate, exhausted in body and spirit, without hope. A glaring light pierces the ceiling to illuminate the wretchedness of the inhabitants, challenging us to witness this spectacle of misery with horrified if compassionate eyes. There is nothing to celebrate, no dignity to acclaim, no hidden agenda, certainly no one to scorn—only a summons to try to imagine what a life bereft of supports that *we* take for granted must have been like.

It is an art of *access,* culture as revelation, not exhilaration; the only resistance it implies is our own reluctance to examine the paralyzing pain of its contents. What could Haas have had in mind but the desire to leave a visual heritage of this pain, since he knew its living exemplars were soon to vanish from the scene? Such paintings speak with a silent eloquence of existence shorn of its roots, surrounded by physical oppression so debilitating that one wonders how its victims could have endured as long as they did. But this is not an invitation to rhetorical speculation about "inner strength" and "spiritual determination" that might transform confrontation into a form of verbal coping. Haas's title, *Expecting the Worst,* to say nothing of his stark and gloomy draftsmanship, makes his intention unmistakable. Look and see, and seeing, grieve.[7]

If we did not know better, we might be tempted to conclude that some of these paintings were actually designed to *mock* an expression like "cultural resistance to genocide." Bedřich Fritta (pseudonym of Fritz Taussig), a friend and contemporary of Leo Haas's in Theresienstadt, murdered by the Germans when they discovered his hidden drawings, left behind one that takes us into the very lazar house of the deathcamp universe. Called *Quarters of the Aged,* it represents a view of three old men on a three-tiered bunk framed neatly by the rounded arch of the barred barrack window. Against the ostensible normalcy of the title, implying some kind of senior citizen housing, Fritta balances three corpse-like, starving figures with skeletal limbs, exposed rib cages, and fleshless faces with sockets for eyes—a vision of unaccommodated creatureliness in the agony of its death throes. his is how the elderly lived in Theresienstadt, says Fritta, drawing us of our safe orbit into the atmosphere of deprivation and uncer- hat governed the lives of the victims.

The power of such works lies not in any statement they make about the human will to survive, or even art's capacity to transcend an annihilating reality through the permanence of form. The subject of Fritta's sketch is the power of Nazi cruelty and indifference to abrogate transcendence and to bar the entry of any spirit to inhibit the withering flesh and the prolonged convulsions of the body's decay.[8] Haas and Fritta use a form of culture to depict the genocidal impulse that was beyond the imagination of a kindly man like Janusz Korczak. His fantasy of an angel of death might ironically have been subtitled *Hoping for the Best.* Haas's *Expecting the Worst* was itself an understatement, since what awaited the victims he portrayed on their arrival in Auschwitz was equally beyond the imagination, though kindness had nothing to do with it.

Why is it that some commentators speak of a Theater of Courage when alluding to cultural efforts like Korczak's? One might just as easily (and accurately) refer to a Theater of Desperation, a poetry of the impossible, what I have elsewhere called an art of atrocity. All of us share the need to find oases of hope amid the deserts of deathcamp destruction. But we know that oases belong to the realm of illusion as well as reality, and distinguishing between the authentic and the imaginary is no simple task. Culture in regard to the Holocaust emerges from the orbit of ordered experience, but mirrors an ordeal that is misshapen, abnormal, inhuman, grotesque. We may stand before a funhouse mirror and laugh, because we know that the distortions are temporary exaggerations which will disappear as soon as we restore our normal selves to the light of normal day. But when life became a nightmare from which men and women did *not* awaken, such transpositions ceased to be available, and the consolations of culture became an uncommon phenomenon that we label a form of resistance only at the risk of distorting its role in the career of victim and survivor.

One of the most extraordinary documentations of the silences *and* choking expressions of culture in the deathcamp universe is Charlotte Delbo's multivolume account of her experiences as a deportee, *Auschwitz and After.*[9] Disciple and associate of Louis Jouvet, famous French theater impresario, Delbo brought with her to Auschwitz an atmosphere of culture whose fate in the camp exposes its limitations and strengths when confronted by the atrocity of genocide. In a remarkable, little-known story called "Phantoms, My Companions," Delbo prepares us for this exposure by telling of her boxcar journey to Auschwitz accompanied by the creatures of the French literary imagination: Julien Sorel, Fabrice del Dongo, Ondine, Phèdre, and many others, but especially Molière's misanthrope, Alceste. They constitute

Delbo's inner security on this voyage to the unknown, her main resources of "resistance" against the threat of genocide. Leaving behind her "annihilated love," her husband, shot by the Germans, she wonders how she will endure her loneliness and vulnerability.

Then suddenly a figure materializes from the darkness of the boxcar: it is Alceste. A strange conversation ensues midway on this voyage from France to Auschwitz, from the realm of normalcy to a circle of atrocity worse than anything Dante imagined. "Alceste, you're traveling with us?" she asks. "With you," he replies. Like us, still emotionally attached to the realm of normalcy, lacking access to the circle of atrocity (where she has not yet arrived), the prisoner is appeased and fortified, almost gladdened: "I found out that Alceste, the hero of a play, was also capable of human heroism, he who was accompanying me by choice where I was constrained to go, forced by force of arms." The others gradually disappear, terrified by the prospect before them, but Alceste remains. For a moment, the barriers separating art from reality disintegrate, so that Delbo, bereft of human consolation, can gain support by entering into an alliance with this literary creation. The process is not so distinct from what Korczak must have hoped for from his orphanage performance. But Delbo is shrewder than Korczak; an artist herself, she understands the limits of her discipline, the boundaries of the imagination, the rift between art and the inhuman.

As the transport draws closer to its destination, Alceste is determined to remain with her; but when it arrives, he leaps down into the icy air, amid barking dogs, shouted commands, and numbed bodies—and disappears. Delbo concludes,

> Alceste had not been able to endure the desolate scenery which met his eye as he came down from the train. . . . Certainly, he had intended to pursue in the desert his ideal of solitude, but, as I had feared in fact, he could not live so far from other men, from the human.

Molière's misanthrope, whose ambition and resolve had been to leave mankind and migrate to the desert, reveals the limits of his literary existence and the cultural milieu that gave birth to him when he is unable to tolerate the desolation that greets his eyes as he leaps from the boxcar in Auschwitz. The gulf dividing art from life reasserts itself here; culture had creative options, as the children in Treblinka may have discovered, that abandoned human beings do not. Alceste's vaunted ideal of solitude has nothing to do with the atrocities of the deathcamps, which would make his moderate brand of misanthropy blush.

Nor have the atrocities of the deathcamps much to do with our ideal of culture, though we are more reluctant to acknowledge the fragile connection than was Alceste. Delbo confirms its fragility when she forgives Alceste for leaving her: "I was all the more grateful at the thought that he had accompanied me this far. . . . I never found him again throughout this sojourn in hell. Nor did I find any of his friends, or of my former friends." She does not repudiate culture, but she learns that the support she will require to "resist" the assaults of Auschwitz will come from the other companions on her transport, the women who "had the solid, fraternal qualities needed for such an unequal fight." In a variation of Jean Améry's truth, she discovered that no bridge led from misanthropy in Molière to the inhuman in Auschwitz. The art she knew could not flourish in such an atmosphere; characters like Alceste "could never have been born in such total misery, in this daily horror of filth and suffering."[10]

The problem was not "cultural resistance to genocide," but resistance to cultural genocide, and seen in this way, our subject assumes a more sanguine aspect. Poems, plays, and musical performances did not preserve human life, but they maintained a continuity in human culture. The art of the concentration camps and deathcamps, whatever form it took, represents an effort on the part of doomed European Jewry to perpetuate the energy of imaginative activity even as their bodies perished. Against the burning of the books, the refusal to perform Jewish composers, the exhibits of so-called decadent painting, the crude anti-Semitic films, some rare Jewish talents found means to portray the features of their doom.

There is little evidence, however, that they believed their art would be effectual against the determined German efforts to annihilate them. We know how ineffectual it was. One might even argue that in their visions the artist-victims of the Holocaust, whom today we celebrate for their cultural resistance to genocide, were vividly aware of how futile art was against the enemies of human life. The hope that their work generates for the durability of human culture simultaneously exposes the ravages of atrocity against being and existence, the human reality that is the substance of but not equivalent to the art that reflects it. "Imagining," Charlotte Delbo writes in one of the volumes of her memoirs, "is the first luxury of an adequately nourished body, enjoying a minimum of free time, having at its disposal a nucleus for fashioning dreams. At Auschwitz, you didn't dream—you were delirious."[11] We need perhaps more often to be reminded than informed that art is not life, that genocide was the foe of life, and culture the friend of art. Janusz Korczak, with the best of intentions, but mistaking art for reality, tried to give his children a living dream;

Treblinka presented them with an irrevocable nightmare. The imagination cannot be blamed for invoking dreams like "cultural resistance" in its desperate wish to escape from deliriums we can scarcely endure. But the deliriums remain, and culture serves us best when it directs our attention to this fact.

That it can do so with emotional intensity and sometimes a dark kind of beauty is a tribute to the resilience of art, rather than to the heroism of men. In a poem called "Burnt Pearls" written in 1943 in the Vilna ghetto (which he survived), distinguished Israeli poet Abraham Sutzkever has left us a powerfully concentrated illustration of the tentative relationship between culture and genocide during that period—but it has nothing to do with resistance:

> It is not just because my words quiver
> like broken hands grasping for aid,
> or that they sharpen themselves
> like teeth on the prowl in darkness,
> that you, written word, substitute for my world,
> flare up the coals of my anger.
>
> It is because your sounds
> glint like burnt pearls
> discovered in an extinguished pyre
> And no one—not even I—shredded by time
> can recognize the woman drenched in flame
> for all that remains of her now
> are these grey pearls
> smouldering in the ash—[12]

This is confrontation driven by an inviolable spirit of authenticity. The broken hands, the sharp teeth, the darkness and anger, the pyre, the flame, the ash, the time-shredded poet himself, and the indistinguishable corpse of the victim, violated by fire—these are the timeless emblems of Holocaust, and language cannot undo the vehemence of their cruelty.

But Sutzkever's poem is also art of the highest order, which would not have existed in its present form had there been no Holocaust. It does not "resist" genocide; it is born of it. It reflects the power of word and image to immortalize anguish: not to celebrate the victim's transfiguration into a glorious example of spiritual defiance, but to mourn verbally her disfiguration, her disintegration, her change from flesh into ash. Culture here is not a collection of shining white pearls dazzling the future with mementos of heroic endurance, but mere burnt grey ones, of precious if diminished value, testimony to an irretrievable loss, imperishable substitutes—but only that—for a world that is no more. If meaning is to be found in such loss, it must be searched for in the tentative bond linking the burnt-pearl remnants of the victim's death with the vital, glinting memories inspired by Sutzkever's words. We need to honor and cherish, but not to distort or misapprehend, their inestimable worth.

1987

6

Understanding Atrocity: Killers and Victims in the Holocaust

he Murders at Bullenhuser Damm is a documentary account, painstakingly assembled by a West German journalist, of the hanging of twenty Jewish children between the ages of five and twelve in an auxiliary camp of Neuengamme concentration camp near Hamburg during the closing days of World War II.[1] The accused killers, SS guards, officers, and doctors, called their activity "obeying orders from above." We would call it "concealing the evidence" of some of the most bizarre and senseless medical experiments ever performed on human subjects in the name of scientific progress during the melancholy years of the Third Reich. The details of the experiments, their consequences, and the fate of the young victims make sad but compelling reading. But the real value of this important book is the insights it gives us into the minds of the killers, ordinary men of no singular ability who gain our attention because they were powerful and willing enough to murder, without compunction or demurral, little Jewish children.

Dr. Kurt Heissmeyer, had there been no war, would probably have spent his life as an obscure country practitioner, ministering to the aches and pains of the local population, perhaps also promoting himself, though without adequate training or ability, as a specialist in diseases of the lung. Heissmeyer picked up the already discredited idea of two Austrian physicians that tuberculosis of the lungs could be combated by artificially inducing tuberculosis of the skin. As a believer in Nazi racial theory, he thought it was not feasible to study the origins of human tuberculosis through experiments on animals, "for such a procedure suggests that the constitution of animals is the same as that of man." Thus Heissmeyer received Himmler's permission to use human subjects—Jewish concentration camp prisoners. He began with adults, but the results were negative, so he turned to children, speculating that they might be more prone to immunity. Twenty Jewish children from Poland, France, Holland, Italy, and Yugoslavia were sent from Auschwitz to Neuengamme. Having failed to develop a serum to heal TB in adults (he discovered this by inoculating them with live tubercle bacilli, then performing autopsies on the bodies of the hanged victims), he sent for the children, to see if they were naturally immune or easier to immunize than their elders.

Although eleven of his SS associates were arrested, tried, convicted, and executed in 1946 for the murder of the Jewish children, Heissmeyer himself was not discovered until 1959, working as a physician in East Germany and running the only private TB clinic in the country. At his trial, he admitted that he never would have tried his experiments on ordinary human beings, though prisoners were another matter: "I did not think inmates of a camp had full value as human beings." Pressed on why he did not use guinea pigs for his research instead, he carelessly replied: "For me there was no basic difference between human beings and guinea pigs," then corrected himself— "Between Jews and guinea pigs." His current patients, apparently untroubled by the accusations against him, testified warmly in his behalf, one of them exclaiming: "He was like a father to those of us who were ill." One wonders what had happened to his paternal instincts as he sought authorization from Berlin to get his young Jewish charges out of the way before the Allies overran Neuengamme. We are not alone in this; after he was sentenced to life imprisonment in 1966, one of Heissmeyer's sons mused: "I am still grappling with all of this: How can a man who loved children do such things?"

Unfortunately, the answer is "Easily," and this book confirms such an unhappy conclusion. At his trial, Heissmeyer confessed that he was unable to feel the remorse that was expected of him, and his reason— or rationalization—illuminates the inner mental process that must have determined the responses of war criminals far less mediocre than Dr. Kurt Heissmeyer: "I expect that if one is to judge fairly, one should weigh the bad deeds against the good, and this might perhaps prove that I helped more patients than I did not." Neither guilt nor contrition afflicts the accused, whose mental gymnastics establishes the surprising moral calculus whereby twenty-one "saved" German patients compensate for the twenty murdered Jewish children. More mystifying is Heissmeyer's conviction that he "did not intend anything bad with these experiments and did not commit anything bad, if one disregards the concentration camp." Clearly, in his thinking traditional distinctions between good and evil never did and still do not apply to the concentration camp universe, since its Jewish victims did not and do not possess "full value as human beings" and hence neither were treated then nor are remembered today as individuals deserving such consideration. Some readers may be relieved to learn that a year after his sentencing, Heissmeyer died in prison of a heart attack.

Equally illuminating is the attitude of Alfred Trzebinski, senior SS physician at Neuengamme and Heissmeyer's supervisor in the medical experiments. At his trial he spoke freely and frankly about the murder of the children. He had been made ill by what he saw, but there was nothing he could have done: "If I had acted as a hero the children might have died a little later, but their fate could no longer be averted." The compelling power of "orders" may have seemed a convincing excuse to Trzebinski at his trial, but it is a feeble one to the reader sensitive to the historical moment of the killings. They occurred on April 20, 1945: the war was long since lost, Germany lay in ruins, the Allies were nearing Neuengamme, in ten days Hitler would commit suicide in his bunker, and a week after that Germany would surrender unconditionally. Was Trzebinski aware of none of this? How could the murder of these twenty children be advantageous to the Third Reich, which no longer had a future? On the other hand, no one seemed averse to the opportunity for eliminating twenty more Jews, and the compelling nature of *this* possibility, in the closing months and days of the war, is too familiar to require documentation.

What impressed Trzebinski most about his behavior was his charity in administering morphine injections to the children before they were hanged. Although he defined the situation accurately by admitting that "you cannot execute children, you can only murder them," this SS doctor did not feel any complicity in the deed. Despite the absence of any attempt to prevent the murders, or of any effort to hinder, delay, or postpone them, he nevertheless could not understand why doctors who "fulfill the unquestionably necessary task of determining death are charged with this as a crime, even in the case of illegal execution." But this is the same man who had confessed that you couldn't execute children, only murder them. With unruffled conscience Trzebinski could testify to his own exemplary conduct, as if such an unspeakable crime might unfold without touching his image of himself as a compassionate and moral human being. Part of the explanation is that the victims were "only" Jews, and most of us still have trouble today understanding how easily this fact transformed atrocity for the Nazi mentality into a necessary and justifiable act. But another part of the explanation has nothing to do with shifting values or altered moralities. It concerns the blurring and decay in Nazi Germany of what we might call the phenomenon of *agency.* It involved the development of a psychology of detachment whereby the perpetrator of a violent act separated himself inwardly from its effect upon the victim, especially (though not exclusively) when the victim was Jewish. It was as if Nazi policy were filtered through a human being while simultaneously discouraging the individual from feeling responsible for what he was doing. The result was an attitude, which survived the war (as Nuremberg and the other trials confirm), of apparently genuine perplexity when men who never perceived themselves as criminals were charged with the most horrifying crimes.

This explains the response of Max Pauly, commandant of Neuengamme, where 55,000 prisoners died, who could not reproach himself for anything he did during the war, including transmitting the order for the murder of twenty Jewish children. "They say everything we did was bad," he writes his son from prison. "It was a concentration camp authorized *by the state,* yet I am nevertheless held responsible for everything that higher-ups ordered me to do." Pauly's willing and proud and enthusiastic compliance with these orders seems irrelevant to him now, so successfully has the "state" dried up in him the phenomenon of agency. His last testament to his son from his death cell affirms the view that Max Pauly had ceased to exist as a private individual, to be replaced by a collective personality that absorbed his being, judged his actions and justified his past existence: "Always be proud of being a German, and detest with all your might those who acquiesced in this absolutely false verdict. . . . As for me, I find great consolation in the fact that the German people as a whole unequivocally reject this pernicious propaganda of lies." Pauly is not guilty of moral depravity, or even, I suspect, of self-deception; he suffers from the illusion, spread by the Nazi system, that one could be the commandant of a concentration camp and still not be personally responsible for the fate of the victims who died there. "I am not aware of any fault of mine," he writes in another letter; "I always acted in the interests of the prisoners, doing my duty to the very end."

The twisted notion that murder could sometimes be in the prisoners' interest is less far-fetched than it sounds, when we consider the verdict of the public prosecutor of Hamburg, who in 1964 was asked to conduct an investigation into the possible culpability of the one SS man involved in the murder of the children who had never been indicted for the crime. The prosecutor ruled that there was insufficient evidence for indictment. His reasons represent the single verbal moment in this sad but significant book that reduced me, for the first time in my twenty-year confrontation with the Holocaust experience, to a stunned and baffled silence:

> The investigations did not prove with the certainty that is demanded of them that the children suffered *unduly* before they died. On the

> contrary, much can be said for the fact that all the children became unconscious as soon as they received the first injection and were therefore not aware of all that happened to them thereafter. *And so, beyond the destruction of their lives no further harm was done to them; and in particular, they did not have to suffer especially long, either in body or soul.* (emphasis added)

We will never understand the crimes of Nazi Germany, or the roots of the Holocaust, until we penetrate the kind of thinking that inspires a human being to regard the destruction of lives with such trifling inconsequence. The legacy, unfortunately, lives on beyond the cellars of Bullenhuser Damm.

Had an intelligent and sensitive SS man tried to keep a diary similar to *An Interrupted Life: The Diaries of Etty Hillesum, 1941–1943*, he would soon have been frustrated and forced to abandon the project.[2] The evolving theme of her journal, appearing in an early entry, would have made no sense to him: "It is a slow and painful process, this striving after true inner freedom." The SS man's identity, insofar as he possessed one, was externally determined. Subjective responses to surrounding reality lost their importance as clues to his personality. "Inner being" ceased to exist as a meaningful concept to those engaged in the routine slaughter of millions. Hence the sullen resentment of so many Germans after the war accused of crimes that they could not recognize as crimes for which they were responsible. They had not violated a nonexistent inner commitment to moral freedom, and they had not violated an external loyalty to a state and a system which had relieved them of the sense of personal responsibility for their actions. Of what were they guilty?

Life was quite otherwise for Etty Hillesum, who created an inner universe simply to survive, as Nazi oppression of the Jews in Holland intensified. If reality did not chime with her imagination, she did not surrender to reality, but let her imagination run riot. She refused to believe that life had no meaning, even if it took a long lifetime to find it. In fact, it took her only a brief and sudden deathtime to find it, though we will never know what meaning she "found" as she stumbled from the boxcar that had brought her to Auschwitz: she did not return. The *Diaries* recount the growth of her mind and her search for identity from March 1941 until September 1943, when she was deported from the Dutch transit camp at Westerbork. Her entries were passionately concerned with her self as a woman, an erotic being, a spiritual creature, and—what I think is the unexpressed secret of the *Diaries*—a latent convert to Christianity.

What are we to make of this twenty-seven-year-old Jewish victim of the Holocaust who sits on her fifty-five-year-old Jewish lover's lap (her other lover is even older) while he reads to her from Thomas à Kempis, who writes that sometimes she longs for a convent cell, who spends more and more of her waking hours praying on her knees (certainly not the traditional Jewish posture for prayer) in quasi-mystical and saintlike dialogue with God, quoting from the Gospel of Matthew and from St. Augustine, and translating from the Russian Dostoevsky's *The Idiot,* a novel whose protagonist resembles a Christ-figure more than any other in that author's fiction? Although she does not repudiate her Jewish identity, neither does she particularly acknowledge it; it lies uneasily on her shoulders like a burden she might wish to discard. Indeed, her *Diaries* are virtually unique among victim or survivor memoirs in dissociating her Jewish heritage and her fate and searching for sources of strength and consolation in a Christian vocabulary and attitude that to many will seem alien to her special dilemma. As the external supports of her life collapse or disappear, as her professors at the university are deported or commit suicide, she retreats—or advances—to a position of spiritual fortitude that will support her until the end. But by writing about the destruction of Dutch Jewry as if it were an example of Christian suffering and martyrdom she obscures, perhaps quite unintentionally, some essential differences between the two experiences. She casts the event into a context which does justice neither to the idea and meaning of Christian suffering nor to the ordeal of European Jewry during the period we have come to call the Holocaust.

There are intimate moments early in the *Diaries* when Etty Hillesum appears as an ordinary and even commonplace young woman. "My immediate reaction on meeting a man," she writes, "is invariably to gauge his sexual possibilities. I recognise this as a bad habit that must be stamped out." The encroaching threat of an unspecific but unprecedented suffering slowly causes her to shift the emphasis of her entries from the promptings of her strong physical nature to a search for religious and spiritual tranquility, but without abandoning her infectious enthusiasm for life. Our hindsight, however, darkens her efforts to reconcile hope with annihilation, making her buoyancy seem pathetic and exasperating rather than praiseworthy. "I have already died a thousand deaths in a thousand concentration camps," she proclaims. "In one way or another I know it all. And yet I find life beautiful and meaningful." Such self-assurance blurs the distinctions between imagined pain and real atrocity, and does little to help the uninformed reader appreciate the precarious situation of the Jews in Holland before their deportation. Her breathtaking naïveté reveals how out of touch she was with the daily fears and doubts of these victims. "But does it matter," she asks, "if it is the Inquisition that causes people to suffer in one century and war and pogroms in another. . . ? Suffering has always been with us, does it really matter in what form it comes?" One can only mutter helplessly today that to the starved and the beaten, to the tortured and the gassed, it must have mattered. She spoke bravely of "our common destiny" and "the story of our fate," as if some power in the universe had designed Jewish history to culminate in Auschwitz and the other deathcamps. It may be unfair to call Etty Hillesum's use of traditional vocabulary like "fate," "suffering," and "destiny" specious—"inappropriate" would be a juster term—but at least we should pursue the direction in which it leads (and misleads) us, in order to perceive more accurately the scenario of her vision.

It may have been Jesus's destiny to die on the cross, and the destiny of his Christian followers to suffer in defense of it, but it never was Jewish destiny to die in the gas chamber, any more than the suffering preceding such a doom was fated for the victims, or the anguish following it for the survivors. Etty Hillesum equated what she called "living out one's destiny" with a metaphor she inherited from the nineteenth century: "I don't have to mess about with my thoughts any more or tinker with my life, for an organic process is at work." One would have thought that the carnage of World War I would have disarmed this metaphor forever; but obviously its inspirational force survived into the middle years of our century, to nurture consolation and confidence in a victim like Etty Hillesum. How else could she have decided that if "one burdens the future with one's worries, it cannot grow organically"? How else could she have concluded that "even when things go badly for me I shall still find life good and worth living"? We have not yet recovered from the trauma of the utter collapse of that organic metaphor, the legacy of a Romantic era, and everything it implies: that no grief is ultimate; that the human capacity for suffering is equal to any anguish; that, in Hillesum's words, "now death has come as large as life and I greet him as an old acquaintance. Everything is so simple."

In a classic formulation, Jean Améry, an Auschwitz survivor, wrote after his return that "no bridge led from death in Auschwitz to *Death in Venice*." He meant that images of dying supported by the pre-Holocaust imagination were cancelled by the unprecedented experience of gas chambers and crematoria. Améry knew what Hillesum had yet to learn: asphyxiation was not Jewish destiny, and cremation was not Jewish fate. What Hillesum *might* have recorded in her diary between her arrival in Auschwitz on September 7, 1943, and her death there on November 30, we will never know. She seems to have had a tiny glimpse of the future when she confessed in her published *Diaries* that what was at stake was "our impending destruction and annihilation, we can have no more illusions about that." But this was never her final position, since she continued to have illusions, writing only a week later about a "medical exemption" from deportation "on account of my inflamed kidneys and bladder"! Her *Diaries* are the record of a mind multiplying possibilities into the future in a way that suggests literary rather than literal imaginings. One is tempted to conclude that she was experimenting with language too, seeking the gnomic expression more than the vividly observed detail: "[B]y excluding death from our life we cannot live a full life, and by admitting death into our life we enlarge and enrich it." If she were speaking of the normal sequence of a life, this might have some force, but she is talking of Auschwitz and annihilation (her own word), and who among us feels "enriched" by their presence in our unhappy century?

Though Etty Hillesum wrestled valiantly with language to find an idiom commensurate with the ordeal of Dutch Jewry, she did not succeed. Her vocabulary was too limited by a tradition of thought reluctant to anticipate what might lie ahead. In this, of course, she was not alone, though others have failed with less certitude and greater humility. A certain arrogance of tone and style will not endear her to some readers. Her verbalized courage approaches rhetoric: "[T]he main thing is that even as we die a terrible death we are able to feel right up to the very last moment that life has meaning and beauty, that we have realized our potential and lived a good life. I can't really put it into words." Up to the very last moment? And how should we have explained that to the Polish mother whose final words to her young daughter as they awaited death in the anteroom to the gas chamber were recorded by Salmen Lewental in a diary dug up in the vicinity of Auschwitz years after the war: "On account of you my pain is so great that I am dying when I think of it." She too couldn't really put "it" into words, but the quiet eloquence of her despair registers more authentically on our pulses than the rhetoric of hope that inspired Etty Hillesum. Hillesum's bland conclusion that "ultimately they cannot rob us of anything that matters" may remind some uncomfortably of the Hamburg prosecutor's conclusion about the hanged Jewish children: "[B]eyond the destruction of their lives no further harm was done to them." Though the circumstances behind these statements differ radically, in both instances language displays a felicitous capacity to shield the speakers from horrors they prefer not to confront. In a moment of stunning clarity, Etty Hillesum veers toward this truth—"I am still talking in much too philosophical, much too bookish a way, as if I had thought it all up to make life more pleasant for myself"—but she does not pursue it, deciding instead to quote with approval her friend's observation that "it is a great privilege, isn't it, that we have been chosen to bear all this?"

We who know better—or should—will experience the powerful irony looming behind virtually every statement Etty Hillesum makes, illuminating the gulf between the idea and the experience of suffering that she scarcely recognized. Only once, when she ceases to be a diarist and sends a long letter from Westerbork transit camp describing the fear, confusion, and helplessness of mothers and children preparing for deportation the following morning to an unknown destination, does she discard her inspirational rhetoric and describe simply and with utter compassion the cruel circumstances dooming these Jewish victims. A different writer emerges here, a vivid and accurate observer who has momentarily turned away from her own spiritual dilemmas and accepted the more modest role of recorder of human wretchedness. This is the Etty Hillesum, the sympathetic "ears and eyes of a piece of Jewish history," as she puts it, one would have wished to hear more from.[3]

1985

7

Fictional Facts and Factual Fictions: History in Holocaust Literature

I call this essay fictional facts and factual fictions in order to suggest a symbiotic kinship between actual and imaginative truth in the literature of the Holocaust. One of the many tasks of Holocaust criticism is to clarify the complex bond, in the minds of both author and audience, linking the oppressions of history to the impressions of art. Normally, the artist is free, and indeed expected, to manipulate reality in any way his vision sees fit. The imagination seizes experience, drops it into its crucible, allows it to ferment, and offers the results to a tolerant and often eager public. Few readers today fault Tolstoy's art for being "unfair" to Napoleon in *War and Peace,* any more than they censure Hawthorne for adapting Puritan values to the needs of his theme in various stories and novels. But Rolf Hochhuth roused a storm of controversy when he dramatized the "facts" of Pope Pius XII's attitude toward the Jews, thus making of it a fictional fact that forced audiences to acknowledge the tension between actual and imagined truth. Similarly, in *King of the Jews,* Leslie Epstein antagonized many by transforming the "facts" of Chaim Rumkowski's leadership in the Lodz ghetto into the extravagant antics of I. C. Trumpelman. Somehow, they felt, the imagined tone of comedy and farce seemed inappropriate to the actual details of the atrocity it sought to illuminate. The same, of course, has been said repeatedly of Lina Wertmüller's film *Seven Beauties,* which offends the memory and sensibilities of viewers who anticipate greater fidelity to the realities of the concentration camp ordeal.

When the Holocaust is the theme, history imposes limitations on the supposed flexibility of artistic license. We are confronted by the perplexing challenge of the reversal of normal creative procedure: instead of Holocaust fictions liberating the facts and expanding the range of their implications, Holocaust facts enclose the fictions, drawing the reader into an ever-narrower area of association, where history and art stand guard over their respective territories, wary of abuses that either may commit upon the other.

The problem is not exclusively the reader's or author's. Essentially, the problem is time's, and eventually time will solve it. What will happen, for example, when the specific details of the atrocities at Babi Yar and Auschwitz are forgotten, when their associations with the Holocaust have passed beyond historical memory and they become mere place-names as obscure to their audiences as Borodino and Tagliamento are to Tolstoy's and Hemingway's readers today? In time, in other words, the boundaries separating the historical moment from its imaginative rendition will be blurred, and it will no longer matter so much whether fictional facts, tied to the actual deeds of history, have become factual fictions, monuments to artistic vision that require no defense or justification, but stand or fall on the strength of their aesthetic mastery of material.

But today it still matters, because the urgency of the historical event continues to exert its mysterious power over modern consciousness. A fictional Napoleon no longer triggers an alarm, but the presence of Rudolf Hoess, commandant of Auschwitz, in William Styron's *Sophie's Choice,* sets off sirens of wary constraint. One admires the imaginative courage, if not the results, of George Steiner's effort to write a fiction whose central fact is that Israeli agents have discovered an aged Adolf Hitler, very much alive, deep in a South American jungle. Reading that novella, *The Portage to San Cristobal of A.H.,* imposes almost impossible burdens on the audience's willing suspension of disbelief, confronted as it is simultaneously with the historical fact of Hitler's death and his artistic resurrection, a burden that would be negligible if Steiner's intention were not so serious and the figure were not so centrally identified with the experience of the Holocaust. By recreating a Holocaust personality whose motives still have not been firmly established (to say nothing of being understood) in *fact,* Steiner establishes a wall of resistance to block the passage leading from history to art. Once the historical Hitler becomes a fictional fact, he inadvertently joins the fraternity of men; then (since his creator eschews melodrama), we are forced to face him in fiction as we were reluctant to do in fact. The factual fiction, in other words, becomes a figure we do not want to confront, partly because the real man is still so much more vivid to our imagination than his invented counterpart, and partly because the fact and the fiction constantly war with each other for higher priority. The Holocaust in fact resists displacement by the Holocaust in fiction, as if the artist were guilty of some unprincipled violation of a sacred shrine. This may help to explain Theodor

Adorno's early cry, constantly repeated by others though later modified by him, that to write poetry after Auschwitz is barbaric.

How may we account for this resistance? If the Holocaust were an event that lent itself to heroic portraiture, or the legendary tone of epic, or even the conventional tale of purification through suffering—and of course much Holocaust literature limits itself to expressing just such traditional themes—then this problem, I think, would not exist. Literature generalizes human experience, while the events of atrocity we call the Holocaust insist on their singularity. The imagination seeks to link the two, to find a bridge through metaphor, image, a language of connection. But the Holocaust has impurified language in a way that prevents this from happening. Norma Rosen quotes a line from a poem by Eugenio Montale—"I've sniffed on the wind the burnt fragrance / of sweet rolls from the ovens"—and suggests that no one today can register those images without some form of associative despair.[1] The catalogue of such terms is long and if they do not deplete available vocabulary, they certainly limit the writer's control of their use: "train," "track," "boxcar," "smoke," "chimney," "ghetto," "roundup," "deport," "roll call," "organize," "camp," "block," "oven," "furnace," "gas," "shower"—even "arrival" and "departure." They are not a lingo (though some unique terms like "kapo" and "musselman" and "Canada" appeared, with many more emerging from the longer Gulag experience that Solzhenitsyn writes of), but ordinary language that normally empowers the writer to give rebirth in literature to the tensions and conflicts of life.

How do we verbalize the enigma of a language that alienates even as it struggles to connect? We can avoid the problem, as many do, by choosing a language of consolation and transforming the Holocaust experience into a story of Jewish resistance and survival, pretending that these terms at least permit the universalization of the ordeal. But those who confront that ordeal with unabashed frankness will have to acknowledge that in its scope it was little more than, and never *will* be much more than, a story of Jewish murder. Since the *factual* fact is so dismal and unedifying, how are we to face the fictional fact that lifts it out of its original bedrock in history (where, four decades later, we still probe for its "meanings" in the layers of time), and translates it into an artificial setting? How can we expect the mind to adjust from the real horror to its portrayal in art, while simultaneously accepting the premise that nothing essential has been lost or changed in the process?

Sometimes I think this is too much to ask, of writer *or* audience. Except for the brief episode near the end of André Schwarz-Bart's *Last of the Just,* I know of no fictional moment set *inside* a gas chamber (though probably others exist). Making a fictional fact of that factual fact raises the specter of the resultant factual fiction's becoming "only" a fictional fiction, by which I mean that for the sake of art, the imagination raises the event to the level of imagined experience and somehow constricts instead of illuminating the implications of that dehumanizing trial. Art in its essence asks us to see life other than it literally was, since all art, even the most objective naturalism, requires selection and composition, and this alters the purity (or in this case the impurity) of the original historical moment. A wholly imagined fictional episode, or one unrelated to a sustained trauma like the Holocaust, faces no such constraints. Hence the temptation is to defer to tradition through heroizing the victim by identifying him or her with familiar forms of suffering and well-known gestures of resistance, or villainizing the persecutor through familiar expressions of cruelty. But these are evasive tactics too.

Another task of Holocaust criticism is to address the implications and consequences of these dilemmas. Some writers are more sensitive to the issue than others. Wishing the life of his heroine in *The White Hotel* to end at Babi Yar, D. M. Thomas—wisely, I think—does not try to reconstruct that episode as a purely imaginative act because he knows, perhaps through instinct, that the Holocaust raises questions of narrative authority few other historical occasions confront us with. Moreover, since the intersection of history with fiction, actual with imaginative truth, is one of the themes of Thomas's novel, introducing a literal eyewitness account of Babi Yar into the text reinforces the issue he is exploring. He takes most of the description of the catastrophe of Babi Yar from Anatoli Kuznetsov's nonfiction novel of he same name. As a young boy, Kuznetsov lived in Kiev and experienced some of the events surrounding the executions at Babi Yar. But even he felt the need for a more reliable narrative authority and, as he tells us, used the account of a survivor, Dina Mironovna Pronicheva. "She is the only eyewitness to come out of it," he says (inaccurately, as it turns out), "and I am now going to tell her story, as I wrote it down from her own words, without adding anything of my own."[2] Curiously, this is a familiar formula to anyone conversant with the history of the novel. Unwilling to be accused of frivolous fabrications, some early novelists habitually fell back on such disclaimers, as if invented narratives would somehow seem less authentic than so-called authoritative testimony. Ironically—for some writers, at any rate—the Holocaust has revived an ancient tradition. Perhaps this accounts for the popularity of one form of Holocaust literature, the documentary novel, represented not only by Kuznetsov's work, but also by Jean-François Steiner's *Treblinka* and Thomas Kenneally's *Schindler's List.*

Thomas's purpose, however, is not to give us an authoritative account of Babi Yar, but to find a fitting dramatic conclusion to the life of his heroine and erstwhile patient of Sigmund Freud, the fictional Lisa Erdmann. The reader, already compelled to identify her character with historical reality by the presence of Freud in the novel, has the merging of her story into fictional fact confirmed by her fate at Babi Yar. "No one could have imagined the scene," she thinks as the executions continue, "because it was happening."[3]

Facts we *know* because they have happened; fictions we only imagine. But the facts of Babi Yar are "unimaginable," and this is why what I call fictional facts play such an important role in our response to and understanding of the Holocaust. By creating an imagined context for Dina Pronicheva's experience at Babi Yar, Thomas makes accessible to the imagination what might have seemed intractable material. But at the same time, he alters the narrative of the real survivor, Dina Pronicheva, whose ordeal is transmitted by Kuzentsov; and in so doing, he creates what I call a factual fiction, since there never was a Lisa Erdmann, so that her "perception" of Babi Yar is an invention. The reader searching for information about or insight into Babi Yar by reading *The White Hotel* runs the same risk as one turning to John Hersey's *The Wall* to learn about the Warsaw ghetto. Art intervenes almost inevitably to mute the impact of history, just as Thomas chooses to end his novel with a fantasy that at the very least qualifies the finality of the anguish represented by the real Babi Yar. Although *The White Hotel* is not exclusively a Holocaust novel, the extermination of the Jews is the denouement of its fictional milieu. The concluding fantasy of transcendence and perhaps of redemption, wholly a product of the imagination, cannot undo the historical horror of the prior episode at Babi Yar. Whether this is the point, or the missed point, of the narrative, each reader must decide in the privacy of his or her critical engagement with the text.

Babi Yar is an episode, a mass execution, perceived through the eyes of a victim, whose own source, we now know, was primarily the testimony of an actual survivor, Dina Pronicheva. In William Styron's *Sophie's Choice* the commandant of Auschwitz, Rudolf Hoess, is a character, with a wife, children, a home, friends, normal desires, while the "episode" of Auschwitz, with its mass executions, forms only the periphery of the narrative. The fictional facts of Hoess's tenure at Auschwitz come from his *Autobiography,* his various war-crimes trial testimonies, and Styron's own visit to the site of the deathcamp. The factual fictions, however, are far more crucial in this work: the motives, the dialogue, the gestures, the portrait not of Commandant Hoess, but of Hoess the private human being, whose primary task almost incidentally happens to be the construction of the gas chambers and crematoria of Auschwitz-Birkenau and the murder of its Jewish inmates. Between 1 and 2 million of them perished there, by his own estimation (though not in the pages of this novel).

Far more challenging than Sophie's manufactured choice—sending her daughter to her death in order to save her son (a decision which any informed reader would know lay beyond the control of Auschwitz's victims, Jewish or not)—far more challenging than Sophie's choiceless choice is *Styron's* choice: to assault our historical consciousness of Hoess the mass murderer with his own conception of Hoess the man, and to expect the resulting fictional facts to be persuasive. If they are not—and I think they are not—the fault lies with both Styron and the tyranny of history, which at this moment in time demands to know not how decent and polite Hoess might have been when he wasn't killing Jews and Poles, but how a man otherwise decent and polite—questionable designations, at best—so effortlessly advanced the cause of genocide.

The factual fictions of Styron's narrative, solely on his own creative authority, would have us believe that Commandant Hoess might apologize to Polish prisoner Sophie for violating his promise by failing to produce her son for her to see. Or that Commandant Hoess, destroyer of Jews and Poles, would not object to the humiliation of vowing to that same Polish prisoner, seriously, not ironically: "You have my assurance and word as a German officer, my word of honor." Styron's boldness in characterizing Hoess matches Steiner's in uncovering Hitler "alive" after all these years, but what fresh understanding of Nazi mentality emerges from their imaginative forays? Is it a revelation that Hoess suffers from migraines; or is exasperated by bureaucratic interference from Berlin; or that he is beholden to former Prussian mannerisms; or admires his Arabian stallion's spontaneity? In attempting to imagine Hoess as someone other than a creature of melodrama, a monster of iniquity, Styron has created an unremarkable figure almost totally dissociated from the deeds that led to his execution after the war. The consequences of those deeds, which engage the reader's consciousness with as much energy as the factual fictions about Hoess in the novel distract it, solicit something more than the familiar outlines of characterization that Styron offers us, something akin to the bizarre contradictions that drive the more complex nature of Nathan.

Literature may not provide sufficient model or inspiration for personages like Rudolf Hoess, but history compounds the dilemma by also hemming in Styron, in spite of himself. He cannot approach Hoess with the same psychological freedom that he does Nathan; yet to do otherwise is to create a wooden creature whose responses in fiction neither illuminate nor are justified by his actions in fact. *Sophie's Choice* is a classic example of the kinship linking the two, constraining instead of freeing the potential for insight inherent in the bond. We cannot say to the reader: take Hoess for what he is, a character in a novel who need not shed light on his real-life prototype. Such a demand would trivialize history, the victims, the critical imagination, the whole macabre enterprise of genocide. But if a character's representational authority does not emerge from the context of the fiction, how are we to respond to him? We return to the intersection of fiction and fact, history and imagination, and that defiant event we call the Holocaust, which the creative faculty will not dismiss but cannot enclose because the limbo it inhabits is shrouded in such uncertain gloom.

Up to this point I have tried to confirm the difficulty, not to say the impossibility, of making a historical figure the center of consciousness in a piece of Holocaust literature. Whether we are thinking of Hitler, Hoess, Chaim Rumkowski, Pope Pius XII, or any other major participant in the catastrophe, the memory of his actual role intrudes on all attempts to dramatize it and erodes the authenticity of their fictional presence. I am convinced that many decades must pass before art will be able to displace memory as the measure of literary success. The most effective Holocaust fictions, like Piotr Rawicz's *Blood from the Sky*, Jorge Semprun's *The Long Voyage*, Tadeusz Borowski's *This Way for the Gas, Ladies and Gentlemen*, Ida Fink's *A Scrap of Time*, and Aharon Appelfeld's various novels, while not ignoring the specific historical context, turn their imaginative beacons on the problem of tone and point of view, angle of vision, centers of consciousness through which the essential atrocity may be filtered. Although fictionalized historical material may alienate the reader, an adequately imagined *invented* center of consciousness can draw him against his will into the net of atrocity, where his own sense of normal reality struggles to escape from the lure. To balance the theoretical discussion in the first part of this essay, I would like to turn now to two texts (one of them, by design, discussed earlier, but with a different emphasis) in order to assess the use of centers of consciousness as bulwarks against the invasion by Holocaust facts of Holocaust fictions.

Although Saul Bellow's *Mr. Sammler's Planet* and Styron's *Sophie's Choice* are not Holocaust novels in the sense of some of the aforementioned works, the title figures in both have had their encounters with mass murder and atrocity. Their lives thereafter have been unalterably affected by their ordeals. Mr. Sammler sees himself as a "*past* person," as "a man who had come back," who "had rejoined life,"[4] though moments of his Holocaust ordeal trickle into the present through his own perceiving consciousness at various points in the narrative. He is his own conduit into the foreground of the novel's action, our only source of authority for the history of the Holocaust "then" and its impact during subsequent years on the mind and vision of Sammler the survivor now.

Sophie is a survivor too, though we shall have to ask how her encounter with atrocity differs from Sammler's. By adopting Stingo as his narrator, Styron prevents Sophie from internalizing her experience as Sammler does, thus excluding the reader from direct access to the inner process, whereby she adjusts the background of her ordeal in Poland and Auschwitz to the foreground of her postwar years in Brooklyn. Sophie offers us only oral testimony; Stingo is the perceiving consciousness, and as an aspiring writer, he seems less interested, his assertions to the contrary notwithstanding, in the "meaning" of Auschwitz than in the literary re-creation of Sophie's story and the gradual revelation of her "secret"—the choice between her children, imposed on her by a Nazi doctor at Auschwitz. Stingo thus duplicates in his own career as a fledgling writer the conflict between fictional facts and factual fictions that beset Styron as author of *Sophie's Choice*. Unfortunately, his efforts to achieve equilibrium are no more successful than his creator's.

The Holocaust background of *Sophie's Choice* is thus Sophie's Holocaust experience, the anguish of the loss of her children, her friends, her health, her self-esteem. That experience evolves slowly into the foreground of the novel through the strict shaping control of Stingo's consciousness, but the evolution is so often disrupted by discordant or farcical concerns that the link joining background to foreground ultimately shatters, never to be repaired. These eccentric matters include Stingo's obsessive fantasies about a final resting place for his wayward organ; Nathan's brutal, lunatic antics; and Sophie's determination to master the intricacies of American, and particularly southern, literary rhetoric. To have a Polish immigrant who mistakes the elementary distinction between seersucker and other kinds of suckers diligently poring over Malcolm Cowley's *Portable Faulkner* or the concise novels of Thomas Wolfe is to strain the reader's credulity. As a writer, Stingo is bound and trapped by his literary commitment to create a portrait of Sophie that is more stylized than human, projecting for us his own version of what a southern American writer might have imagined Sophie to be. The novel is less about Sophie's experience in Auschwitz and her encounter with Commandant Hoess than it is about Stingo's unflagging determination to make literature out of them.

Whether Styron had serious or parodistic intentions with Stingo, the results, for readers interested in the uses of Holocaust facts in fiction, are the same. Striving for literary effects turns out to be incompatible with Sophie's choice between her children, improbable as that was, forced on Sophie by the equally improbable Dr. Jemand von Niemand. But Stingo is unable to restrain his impulse to stylistic embellishment. He taints the purity of Sophie's already melodramatic scream, "I can't choose! I can't choose!" with his private commentary that "tormented angels never screamed so loudly above hell's pandemonium."[5] The compulsion to transform painful human moments into cosmic gestures through stylistic effects like this one makes the Holocaust a gratuitous literary event to be "written up" by an aspiring novelist like Stingo. This *separates* the history from the fiction by advertising the effort to absorb one into the other.

One other example should suffice to make this point. As Sophie's "battered memory" struggles for the last time to face the implications of her children's fate, Stingo again buries her efforts beneath the writer's mound of rhetoric: "She paused to look out through the night at the dark shore of the Virginia of our destination, removed by staggering dimensions of time and space from her own benighted, cursed, and—to me even at that moment—incomprehensible history" (495). One purpose of literary art, especially when confronting the Holocaust, is to encourage a perceiving consciousness to make comprehensible the incomprehensible, but Stingo's penchant for inflated style intervenes repeatedly at crucial potential moments of recognition like these, diverting the reader's attention from Sophie's trial to the writer's resolve to *render* it.

One might expect that Stingo, having extracted from Sophie her ultimate painful revelation, would have devoted himself to sympathy for the victim. But though Sophie's heart may have turned to stone, her body, to Stingo's delight, has retained its sumptuous sinuosity. Indeed, his failure as a perceiving consciousness is that he mistakes a climax for a denouement. On the page following Sophie's confession about her children's fate, Stingo's curiosity about her story is replaced by a night of erotic calisthenics that make one wonder where the focus of his interest has lain all along. What are we to make of this narrator, to say nothing of his narrative, who can literally juxtapose death in the gas chamber with lavish descriptions of the varieties of his sexual experience with Sophie? One longs to believe that here, too, Styron is indulging in parody; but once more, whether he is or not, the results are the same: to mock, to diminish, to negate the authenticity of Sophie's anguish in Auschwitz. Indeed, Stingo parodies Sophie and himself when he describes his last spasm of pleasure in Sophie's mouth in language reminiscent of her response to Dr. Jemand von Niemand's cruel challenge to her to choose between her children. Stingo says of himself: "I verged on a scream, or a prayer, and my vision went blank, and I gratefully perished" (498). So close to her Auschwitz narrative, such evocation of Sophie's language savors of the indecent. Her "oral testimony" itself has collapsed into a bizarre form of sexual parody, as speech disintegrates into suction.

In attempting to write a serious novel about the Holocaust using Stingo as narrator, Styron created for himself insuperable obstacles. Stingo's priapic sensibility, his obsession with tumescence, combined with his platitudinous anti-Semitic instincts, disqualify him for the literary task he chooses—to tell Sophie's story convincingly and sympathetically. The evidence is strewn throughout the narrative. Though Styron may have been indulging in some ethnic "humor" by having his narrator endlessly identify Jews with bagels, Manischewitz, and halvah, or longing to bury his nose in Leslie Lapidus's "damp Jewish bosom" (125), or feeling pleased with the clever turn of phrase that alludes to "window-shopping Hadassah matrons," or asking regretfully "Why, instead of the floundering, broke, unpublished writer that I was, couldn't I be an attractive, intelligent, well-paid Jewish urologist with a sexy wife?" (426)—whatever Styron's intentions, such tiresome facility for stereotyping eventually undermines Stingo's qualifications for the serious business of narrating and interpreting Sophie's Auschwitz background. And a parody of Sophie's ordeal would not be worth telling—certainly not worthy of Styron's talent as a writer. However Styron feels personally about these matters, his real choice of Stingo as perceiving consciousness proves fatal to our appreciation of Sophie's fictional choice in the novel. Stingo's failure betrays Styron's—and vice versa.

Saul Bellow will have none of the Romantic sentimentalism that enables Stingo to speak of SS doctor Jemand von Niemand as "a failed believer seeking redemption, groping for renewed faith" (485). Unlike Stingo, who magnifies the drama of Sophie's ordeal at every opportunity, Mr. Sammler is modest about what he has endured. It was not, he thinks, an achievement: "There was no special merit, there was no wizardry. There was only suffocation escaped" (273). The search for some intellectual, emotional, and spiritual "air" to liberate Sammler's consciousness from this stifling heritage constitutes the burden of the novel, and if his quest is not entirely successful, the fault is history's, not his. History, in the form of his Holocaust encounter with a mass grave, has estranged Sammler from his earlier enthusiasm for the utopian vision of H. G. Wells; it has also alienated him from the student generation of the sixties which hoots him off the platform at Columbia, unaware of the genesis of his present condition of mind and hence totally unsympathetic to his position. Sammler is able to bring his Holocaust experience into the foreground of his *own* consciousness, and he does this throughout the novel; but he can bring it into no one else's, so must bear his memories and their consequences alone.

As a survivor, Sammler thinks near the end of the narrative, "he still has consciousness, earthliness, human actuality" (273). But instead of gaining him entrance to the community of men, these qualities isolate Sammler, since they issue from an exclusive background that he cannot share with others. Sammler consistently deflates the heroic and dramatic possibilities of his encounter with the pit, going so far as to suggest that only chance preserved his life: "If there had been another foot of dirt," he thinks. "Perhaps others *had* been buried alive in that ditch" (273). But this is not the stuff of Romantic defiance and energy that the younger generation around Sammler thrives on; it offers a view diametrically opposed to the enthusiastic schemes of his daughter, his nephew, and his former son-in-law, schemes heedless of the ordeal Sammler has survived. *They* still dream of molding the future; Sammler has been shaped by his past.

And while he has not lapsed into the pessimism of his namesake, Arthur Schopenhauer, Mr. Sammler speaks of the "luxury of nonintimidation by doom" (134) as if he were fighting a rearguard action against the gloomy momentum of modern history. Between the pickpocket's phallic assertiveness against Sammler in the novel's opening episode (a sour if unintended commentary on Stingo's adolescent faith in the power of tumescence) and Eisen's gratuitous violence against the same culprit near the novel's end, Sammler gathers his musings and meditations to find some small justification for his belief that the earth might still be, as he calls it, a "glorious planet" (135). This is in spite of his simultaneous perception that everything was being done by its inhabitants "to make it intolerable to abide here, an unconscious collaboration of all souls spreading madness and poison" (135). Because he himself has had to kill in order to survive, because he has had impressed on him not only the idea but also the experience of life as a mausoleum, Sammler develops internal resistance to the threat of becoming a "bad joke of the self," a threat that both Sophie and Stingo (and Styron?) succumb to when they allow sexual acrobatics to displace sober confrontation with the possibility of the universe as mausoleum. Bellow excludes from his vision the incongruity, not to say the inconsistency of character, that Styron incorporates, thus acknowledging a continuity between Holocaust past and post-Holocaust present that Stingo seems immune to. From the "harsh surgery" of immersion in mass dying, Sammler reflects, one "cannot come out intact" (23). "I assume I am one of you," he volunteers to his audience, "but also I am not" (23). This dual vision constitutes the essence of his perception, what we might call his "good eye" and "bad eye" seeing, the fusion of background and foreground that characterizes his peculiar point of view.

In the end, the main difference between Stingo and Sammler is that Stingo is unable to avoid experiencing Sophie's encounter with atrocity as a writer rather than as a fellow human being. Sammler, on the other hand, has endured the Holocaust in all his creatureliness; his human actuality, as he thinks, is a blend of earthliness and consciousness. Stingo reserves his own encounter with the symbolic grave for the closing paragraphs of *Sophie's Choice,* but his allusion gives away the literary inspiration that has been guiding his investigation from the beginning. Falling asleep on the beach at Coney Island, he tells us, he had "abominable dreams—which seemed to be a compendium of all the tales of Edgar Allan Poe" (515). One would think that Sophie's experience might be enough to shape his nightmares. He views reality through his literary heritage: "All night long I had the sensation of helplessness, speechlessness, an inability to move or cry out against the inexorable weight of the earth as it was flung in *thud-thud-thud*ing rhythm against my rigidly paralyzed, supine body, a living cadaver being prepared for burial in the sands of Egypt" (515).

What Styron has given us in Stingo is the immense, overwhelming egoism of the literary imagination. Sammler condenses into two words—"suffocation escaped"—the ponderous distance between living cadavers in the landscape of Poe and his own escape from "the sand clay and stones of Poland" (273). Stingo's last legacy, to no one's surprise, is some lines from a poem, acknowledging his dream of death but affirming nonetheless *"in glory, the bright, the morning star"* (515). The novel's last words, "excellent and fair," are also from a poem—this time by Emily Dickinson. Are we to assume that literary vision so easily displaces Holocaust fact in the foreground of our imagination?

Once again, Mr. Sammler's concluding aspirations are more modest. The scene of his life has shifted from a mass grave and a mausoleum to a morgue, where the body of his friend Elya Gruner lies in the repose of death. For Sammler, life has become a post-mortem on the implications of his survival; so perhaps it is fitting that his last reflections should be on the fate that links all human beings—the corpse Elya Gruner mirrors our common destiny. Sammler's last gesture of tribute and commemoration is not literary, but human, humbling rather than assertive, even self-effacing: it is the legacy of an aging man, even as the Holocaust is the legacy of an aging planet. What Sammler "knows" in the novel's closing words is, as he had thought earlier, "to be so powerless was death" (289). He does indeed see himself as a "*past* person," someone "between the human and the non-human states, between content and emptiness, between full and void, meaning and not-meaning, between this world and no world" (290). He meditates on the limits within which life on the post-Holocaust planet fluctuates, a shrinking area where Mr. Sammler's melancholy memories leave some space for the decently human and the humanly decent. Holocaust fact and literary fiction share the constraints of these limits. The novels of Bellow and Styron reflect the difficulty of maintaining an equilibrium between them.

1990

8

The Literature of Auschwitz

n the beginning was the testimony. Any study of the literature of Auschwitz may start with judicial versions of the way it was, based on eyewitness accounts—but it can never end there. Readers of the proceedings of the so-called Auschwitz trial of former camp guards that ran for twenty months in Frankfurt (and which I attended for a few days in the summer of 1964) will discover not a narrative leading to insight and understanding, but a futile dispute between accusers and accused. The prisoners in the dock deny virtually everything: Mulka was not there when the prosecution says he was; Kaduk never shot anyone; Boger used his infamous torture instrument, the "Boger swing," only on rare occasions; Klehr was on leave during Christmas 1942, when he was charged with murdering inmates by injecting phenol into the heart. Guilt exists, but the agent is always someone else.

Little in this bizarre courtroom drama leads to a unified vision of the place we call Auschwitz. Scenes remain episodic and anecdotal; scenarios never coalesce; characters stay vague, as protagonists dissolve into helpless victims (through no failure of will, to be sure, but the tyranny of circumstance) while antagonists collapse into mistaken identities or innocent puppets maneuvered from afar. What awareness can emerge from an incident like the following, which casts its net of censure so widely that we strain to unsnarl the agents from the victims?

A group of prisoners in Auschwitz was excavating a ditch filled with water. A witness testifies that some SS men forced them to leap into the ditch:

> They had to jump into the water and swim. Then they ordered a prisoner named Isaac—he was called Isaac the Strong in the camp—to drown his comrades. Finally they also ordered him to kill his own father. In the act of drowning his father, Isaac went berserk and started to scream. So Stark [one of the SS guards on trial] shot Isaac in the water.

Faced with the accusation, Stark spurns his role: "I have nothing to say, because I was not present at any of the incidents the witness described."[1] Such painful testimony may carry the conviction of truth, despite Stark's dissent, but by its very nature the judicial process does not allow it to do so unchallenged. The defense promptly begins to discredit the witness with witnesses of its own, whose evidence may indeed be false—the court is dubious—but how can we ever know? At the moment when Isaac the Strong became Isaac the Mad, he crossed a frontier separating the "normal world" from the abnormal universe of Auschwitz, leaving us beyond the barrier musing on the chance of ever entering into its reality ourselves.

The literature of Auschwitz exists to help us navigate that voyage. It is a perilous journey but a crucial one, if we are ever to admit how little the idea of justice helps us in our efforts to pierce the dark core of that death camp experience. We must confront it on *its* terms, not ours, leaving behind traditional casts of characters with their Isaacs the Strong, and the heroism and tragedy implicit in such titles. In our search for the *meaning* of Auschwitz, to our dismay, we meet often only its absence; what we have to forgo to establish contact with such barren terrain is the theme that absorbs most writers who venture into it.

Not all commentators, of course, agree that the terrain is so barren or that we must forgo so much in order to wander there. The leading exponent of the view that in spite of Auschwitz, life and suffering are unconditionally meaningful, is Viktor Frankl, whose *Man's Search for Meaning* is still probably the most widely read text on the subject. Frankl's strategy is to minimize the atrocities he himself survived, and to stress the connections between pre- and post-Auschwitz reality. Unintentionally confirming the wish of his persecutors, he leaves his fellow victims anonymous, while naming and quoting instead a long list of explorers of the human spiritual condition, including Spinoza, Schopenhauer, Tolstoy, Dostoevsky, Rilke, Nietzsche, and Thomas Mann. The reader is thus prompted to believe, for example, that Tolstoy's Christian novel *Resurrection* or Dostoevsky's Christian declaration—"There is only one thing that I dread: not to be worthy of my sufferings"[2]—is somehow relevant to the Jewish victims of Auschwitz.

Frankl has managed to transform his ordeal in Auschwitz into a renewed encounter with the literary and philosophical giants who preceded its emergence, thus preserving the intellectual and spiritual traditions they championed and his own legacy as an heir to their minds. He uncovers truth by assertion, not analysis, as if the word were an eternally valid bulwark against the dehumanizing assaults of physical violence. Many witnesses in the courtroom at Frankfurt would have been bewildered by the scriptural finality of his proclamation that "if there is a meaning in life at all, then there must be a meaning in suffering."[3]

Frankl's language invites, indeed *requires* us to dismiss the petitions of despair before we confront them. Consider the testimony of Joseph Glück, who happened to be in the witness stand on one of the days that I was present in the courtroom. I remember the figure of a shrunken, elderly Jewish man who seemed crushed and exhausted by his memories, though he had not yet begun to recite them. He whispered hesitantly, seemingly intimidated by the large hall where the proceedings had been moved when the room in the Palace of Justice proved too small, by the ranks of the accused and their lawyers, and by the crowd of spectators.

He had been deported to Auschwitz from Klausenburg with 2800 other Jews, 400 of whom were chosen for work upon arrival. The rest were sent directly to the gas chambers, including his wife, two children, mother, sister with her two children, brother, mother-in-law, and sister-in-law. His situation is certainly not unique, but his statements stun the audience. Asked if he is the sole survivor, he replies "Yes": "For a moment the word hovers over the courtroom, irrevocable but uncertain to whom and where it should turn so that it might not only be heard but also comprehended. The old man sits motionless."[4] No one is foolish enough to ask him if there is meaning in his suffering, or whether he feels worthy of it. The judge shuffles the papers in front of him. Everyone there, including me, grapples with the question of how to translate his simple affirmative, a word haunting the air like a scrap of animated anguish, into a shareable experience. That instant remains vivid to me because it was one of the first times that I asked myself how a literature of Auschwitz, and of the Holocaust in general, might ever achieve such a goal.

This is the focus of our inquiry. From the multiple talents that have addressed themselves to this issue emerges a complex, at times contradictory vision of a way of existing that continues to elude precise definition. For example, against Viktor Frankl's insistence on the power of literature and philosophy to sustain the inner self during the camp ordeal, we have the contrary view of Jean Améry, a classic exposition of the futility of literary memory once it entered the precincts of Auschwitz. Because the mind in Auschwitz collided not merely with death, but with the kind of dying peculiar to an extermination camp, normal responses lost their value. The intellectual was left suddenly defenseless: "Death lay before him, and in him the spirit was still stirring; the latter confronted the former and tried—in vain, to say it straight off—to exemplify its dignity."[5] Frankl argues that Auschwitz challenged the individual to rise above his outward fate, furnishing him "with the chance of achieving something through his own suffering."[6] Améry insists that both fate and suffering disappeared from the vocabulary of Auschwitz, as did death itself, to be replaced by the single fear, shared by all, of *how* one would die: "Dying was omnipresent; death vanished from sight."[7]

Foreseeing the bleakness of a future without the heritage of his literary past, Frankl seems to have decided to treat Auschwitz as a temporary anomaly rather than a permanent rupture; his pledge to inner freedom and the life of the spirit flows from that choice. Entering the world of Auschwitz, as he reflected on it afterward, left Améry with a totally different vista of past *and* future:

> The first result was always the total collapse of the *esthetic* view of death. What I am saying is familiar [though less so, one suspects, than Améry believed]. The intellectual, and especially the intellectual of German education and culture, bears the esthetic view of death within him. It was his legacy from the distant past, at the very latest from the time of German romanticism. It can be more or less characterized by the names Novalis, Schopenhauer, Wagner, and Thomas Mann. For death in its literary, philosophic, or musical form there was no place in Auschwitz. No bridge led from death in Auschwitz to *Death in Venice*. Every poetic evocation of death became intolerable, whether it was Hesse's "Dear Brother Death" or that of Rilke, who said: "Oh Lord, give each his own death." The esthetic view of death had revealed itself to the intellectual as part of an esthetic *mode of life;* where the latter had been all but forgotten, the former was nothing but an elegant trifle. In the camp no Tristan music accompanied death, only the roaring of the SS and the Kapos.[8]

A major function of the literature of Auschwitz is to help us discard the moral, philosophical, and literary systems created by what Améry calls the "esthetic mode of life," systems defining character and action and the tragic sense itself. This becomes especially necessary when they bar the world of Auschwitz from our efforts to enter its realm.

How are we to understand Améry's avowal that in Auschwitz the intellect "nullified itself when at almost every step it ran into its uncrossable borders"? The results were devastating: "The axes of its traditional frames of reference then shattered. Beauty: that was an illusion. Knowledge: that turned out to be a game with ideas."[9] In the presence of this warning, we struggle to prevent our own discourse about Auschwitz from becoming merely a game with ideas. One way out of the dilemma is to accept the threatening possibility of how easily Auschwitz reduced earnest pleas like Rilke's "Oh Lord, give each his own death" to nothing but elegant trifles, utterly irrelevant to the modes of survival available in the camp. Our reluctant surrender of what Améry calls the "esthetic mode of life" is an admission of the powerful role it plays in shaping the conduct and belief of Western civilization and of how bereft we appear without its support, and of the social and economic "modes" that accompany it.

The arriver at Auschwitz entered a world of sensation, not mind: the roar of the SS and Kapos; the bursts of flame and smoke spiraling from chimneys; the rank smell of charred flesh. In a remarkable, little-known story called "Phantoms, My Companions," Auschwitz survivor Charlotte Delbo enacts this encounter, as her narrator journeys by boxcar toward Auschwitz, accompanied by leading characters of the French dramatic repertoire, from Molière to Giradoux. One by one, frightened by the uncertainties looming before them, they leap from the train, until when it arrives and the doors are thrown open, only Alceste, the Misanthrope, remains. He gazes wildly at the bleak, obscure landscape—and disappears, leaving the narrator to the world of corrosive sensation that the aesthetic mode of life had done nothing to prepare Alceste for. Denied the options of art, the narrator faces her doom alone.

Dialogue is the heart of drama; Molière's Misanthrope is disgusted by the social pretensions and hypocrisy that pass for talk among his peers. As Améry describes the content of dialogues in Auschwitz, however, we begin to glimpse how thoroughly camp reality disarmed traditional notions of the aesthetic mode of life, how useless an Alceste would be to counsel or rebuke in such a place: "Inmates carried on conversations about how long it probably takes for the gas in the gas chambers to do its job One speculated on the painfulness of death by phenol injection. Were you to wish yourself a blow to the skull or a slow death through exhaustion in the infirmary?"[10] The literature of Auschwitz makes constant inroads on our assumptions not only about the aesthetic mode of life (which the quoted passage reduces to incoherence), but also about the actuality on which it is based. The universe of dying that was Auschwitz yearns for a language purified of the taint of normality.

Primo Levi, who more than anyone else wrote and rewrote the experience of Auschwitz in search of its significance for him and for us, evokes that instant when at war's end the world of the living faced in embarrassed silence the world of the surviving dead. He tries to analyze the internal content of that momentous meeting, as the first four

Russian soldiers to enter Auschwitz gaze at Levi and his gaunt comrades:

> They did not greet us, nor did they smile; they seemed oppressed not only by compassion, but by a confused restraint, which sealed their lips and bound their eyes to the funereal scene. It was that shame we knew so well, the shame that drowned us after the selections, and every time we had to watch, or submit to, some outrage: the shame the Germans did not know, that the just man experiences at another man's crime; the feeling of guilt that such a crime should exist, that it should have been introduced irrevocably into the world of things that exist, and that his will for good should have proved too weak or null, and should not have availed in defence.[11]

Inversions cancel meaning here, and then challenge its rebirth in the desolate and arid moral soil of Auschwitz. Liberators "oppressed" by compassion; victims shamed by the crimes they witnessed, but did not commit; the innocent feeling guilty, the criminal unashamed; but most of all, the visible failure of good to carry out its historic mission of unmasking and overwhelming evil—such inversions discredit the traditional power of language and the meaning it is accustomed to serve.

More than any other commentator, Levi spent his life trying to explain the nature of the contamination that was Auschwitz. It represented a stain not just on individuals, but on time and history too. The Auschwitz trial was concerned with what was done, and by whom. The premise behind it was that identifying guilt and punishing crime would make a difference in a society based on order. But for Levi, the reach of the moral chaos that nurtured Auschwitz was so vast that justice could not begin to embrace it or define its limits. That was a task for the writer, as Levi shows:

> So for us even the hour of liberty rang out grave and muffled, and filled our souls with joy and yet with a painful sense of pudency, so that we should have liked to wash our consciences and our memories clean from the foulness that lay upon them; and also with anguish because we felt that this should never happen, that now nothing could ever happen good and pure enough to rub out our past, and that the scars of the outrage would remain within us for ever, and in the memories of those who saw it, and in the places where it occurred, and in the stories that we should tell of it.[12]

Levi is honest enough to concede that moral fatigue at the hour of liberty may have been the initial source of this daunting vision. But when he wrote about it soon after war's end, he found the language to confirm its deep and permanent impact on our time:

> Because, and this is the awful privilege of our generation and of my people, no one better than us has ever been able to grasp the incurable nature of the offence, that spreads like a contagion. It is foolish to think that human justice can eradicate it. It is an inexhaustible fount of evil; it breaks the body and the spirit of the submerged, it stifles them and renders them abject; it returns as ignominy upon the oppressors, it perpetuates itself as hatred among the survivors, and swarms around in a thousand ways against the very will of all, as a thirst for revenge, as a moral capitulation, as denial, as weariness, as renunciation.[13]

Some literature of Auschwitz, in a desperate retreat from charges like these about the infection spreading from the very existence of the place, seeks vindication in a countervision that would restore moral health to the victims, imposing shame and accusation on the culprits alone. It resists Levi's acute but paradoxical sense that shame and *self*-accusation, however unwarranted, nonetheless remained a burden for him and many of his fellow former prisoners throughout their lives—in the end, a burden that consumed and probably destroyed him.

The unendurable truths of Auschwitz that Levi expounded with such courage and distinction do not merge easily with traditional literary forms. Thirty years ago, Rolf Hochhuth's bold historical drama *The Deputy* burst on the literary scene in a scandal of controversy. Today it reads like a tame piece of theater indeed, though when I saw it in Vienna in 1964, Hochhuth's setting of the last act in Auschwitz itself seemed a dazzling and agonizing innovation. It testified to how little the imagination must have been prepared then for a literature of Auschwitz. Hochhuth himself paid tribute to this idea in some preliminary remarks in the printed text, revealing his conviction—common at the time, but less so, one hopes, today—that we "lack the imaginative faculties to be able to envision Auschwitz."[14] So much commentary currently exists on the camp (including this volume), that such an attitude now appears naïve and even antiquated. But it was a protective device, one thankfully discarded by artists greater than Hochhuth like Tadeusz Borowski and Charlotte Delbo.

In his efforts to realize Auschwitz on the stage, Hochhuth dismisses documentary naturalism as a stylistic principle, concluding that "no matter how closely we adhere to historical facts, the speech, scene, and events on the stage will be altogether surrealistic."[15] This is a perfectly legitimate point of view; Elie Wiesel describes his arrival at

Auschwitz amid the flames of burning pits as a visionary nightmare. But when Hochhuth shifts from setting to character, he lapses into figures so conventional that he sacrifices any surrealistic effect achieved by the vague movements of the doomed toward the gas chambers in the dim recesses of the stage. The Priest and the Doctor are so clearly defined as adversaries in the contest between Good and Evil that they dwindle into allegory, a finale fatal to *any* adequate representation of Auschwitz in art.

The confrontation between the Priest and the Doctor reflects, on a smaller scale, the epic encounters between Milton's Satan and God's angelic hosts, though Auschwitz has altered the theological balance. The Priest becomes a tragic martyr, improbably choosing to share the fate of the Jews, while the Doctor, the spirit of cynical negation, betrays his ancestry in Ivan Karamazov's devil, a less sinister but equally contemptuous literary prototype. His obvious real model is Josef Mengele, though Hochhuth is not interested in developing a reliable portrait of that notorious figure. He succumbs instead to the blandishments of literary precedents, offering a familiar metaphysical dispute between the ruthless nihilism of the Doctor and the compassion of the Priest, weakened by the failure of the pope and Catholic hierarchy to assert their spiritual force in opposition to the powers of destruction.

Sounding more like Dostoevsky's Grand Inquisitor than a character from a play about Auschwitz, the Doctor unwittingly reveals his literary origins:

> The truth is, Auschwitz refutes
> creator, creation, and the creature.
> Life as an idea is dead.
> This may well be the beginning
> of a great new era,
> a redemption from suffering.
> From this point of view only one crime
> remains: cursed be he who creates life.
> I cremate life.[16]

The commonplace disguised as the profound may betray the limitations of the Doctor's intellect, since his voice falls far short of Ivan Karamazov's mighty indictment of God's world that still beguiles readers into accepting it as Dostoevsky's position too. The real point of *The Deputy* is to charge the Roman Catholic Church with a spiritual timidity that was partly responsible for enabling the historic designs of the Third Reich to end in the gas chambers of Auschwitz. But the verbal premises on which the structure of the play is built—good, evil, conscience, truth, spirit love—do not carry us very far into the daily tensions and moral conflicts of life in Auschwitz itself, nor do they illuminate the natures of men like Mengele, whose elusive motives are hidden behind the taunting, caustic façade of the Doctor's single-minded voice.

A play that discards the metaphysics of Auschwitz as a theme and turns instead to what we might call its materiality is Peter Weiss's *The Investigation*, whose content derives almost wholly from testimony at the trial in Frankfurt. By blending and shaping statements from witnesses and the accused into a carefully organized pattern, Weiss creates from the futile courtroom dispute a fresh vision of the clash in Auschwitz between moral space and destructive place. Although naked testimony may lead the imagination only into confusion, silence, and despair, the form imposed on it by Weiss achieves the opposite effects. Guards and officers in the end indict themselves by their own relentless but transparently false denials of complicity, while the language of the victims finally sheds a terrifying light on the ordeals they succumbed to—or wretchedly survived.

Weiss's drama confirms the difficulty, not to say impossibility, of creating a literature of Auschwitz by relying purely on the powers of invention. In the cast of characters, he identifies each of the "accused" in his play by their real names, since it is these particular agents of mass murder we need to understand, not their dramatic prototypes. The witnesses, however, remain anonymous, spokesmen and spokeswomen for the vast numbers who are unable to speak for themselves (though a few of the witnesses represent the voices of former SS men not on trial who worked in various administrative capacities in the camp). The imagination is drawn into the landscape of Auschwitz by states of feeling inspired by the testimony, condensed evidence that slowly moves us from a sense of how things were to an encounter with what they implied.

Unlike Hochhuth, Weiss does not present Auschwitz as a harrowing inferno, alienating us through the awesomeness of the atrocities committed there. One of his witnesses insists:

> We must drop the lofty view
> that the camp world
> is incomprehensible to us
> We all knew the society
> that produced a government
> capable of creating such camps

Earlier, the same witness had insisted:

> When we talk of our experience nowadays
> with people who were never in a camp
> there is always something
> inconceivable to them about it
> And yet they are the same people
> who in the camp were prisoners and guards[17]

Weiss lowers the barriers of the unimaginable, however, not merely by the statements of his witnesses, but chiefly by the studied arrangement of the eleven multipart "songs" that constitute the scenes of his drama. Beginning with the "Song of the Platform" and the "Song of the Camp" and ending with the "Song of Cyklon B" (the chemical agent of extermination) and the "Song of the Fire Ovens" (the site of physical annihilation), Weiss gradually narrows the space separating the imagination from the camp, leading us from the ramp to the barrack, through various execution sites like the Black Wall (outside) and the cells of Block 11 (indoors), to the gas chambers and the body's final confined destination, the crematorium. The victim's shrinking fate is thus duplicated by the sequence of the testimonies, which in their quest for literal truth have made available through the shaping pen of the dramatist the imagined truth of Auschwitz too.

Unfortunately, the literature of Auschwitz can also be used for political purposes. One distressing enigma of Weiss's text is his refusal to identify the Jews as the primary victims of the murder machinery in the camp, though attentive readers could not possibly mistake his references to the 6 million "persecuted" or to those "killed for racial reasons." Because the events of Auschwitz are still anchored firmly in historical memory, mention of Mengele and Cyklon B and the crematorium are enough to remind us of the destruction of European Jewry. Weiss's universalizing tendency, however, may become more of a problem for future generations, for whom the allusive power of these brief labels will have lost their specific, not to say metaphorical, value. Fortunately, Weiss's play will not be the only source of Holocaust actuality for those generations, for whom the literature of Auschwitz will consist, as it does for us, of a multiplicity of voices and points of view to guide us through its dismal labyrinth.

The boundaries separating the historical moment from its imaginative portrayal may be instructively studied in Elie Wiesel's *Night*, still one of the most concisely powerful narratives of the Auschwitz experience. Although widely read as an autobiographical memoir, *Night* also continues to be classified and critically acclaimed as a novel, and not without reason. Because it is a written text, *Night* suffers the curbs and enjoys the privileges of art, from which courtroom testimony, or any oral account of the Auschwitz ordeal, is exempt. Art in its essence invites us to see life other than it literally was, since all art, even the most objective naturalism, requires selection and composition, thus altering the purity (or, in this instance, the impurity) of the original event. In one of its aspects, Wiesel's text is a study of fathers and sons in Auschwitz, with all inmates being the children of God the Father. This lifts the narrative, to its credit, to be sure, beyond the constraints of autobiography into the realm of imagined fiction; nothing is more "literary" or stylized in the story than the young boy's denunciation of God's world and implied renunciation of its Creator, the seeds of both of which are nurtured by passages in Dostoevsky and Camus, in addition to the conditions of Auschwitz itself.

The literature of Auschwitz is thus bound by its historical context in a way that most other literature is not. Within the above-mentioned constraints, it faces the challenge familiar to all serious writers; finding an appropriate tone and point of view, a suitable angle of vision, a valid and convincing center of consciousness through which to filter the trial of atrocity. Although extravagantly fictionalized historical material about the Holocaust (as in Leslie Epstein's *King of the Jews*) may alienate some readers, a subtly imagined center of consciousness, invented or not, can draw them against their will into the net of human abuse, where their own sense of normal reality struggles to escape from the lure.

One of the finest examples we have of such a strategy is Tadeusz Borowski's collection *This Way for the Gas, Ladies and Gentlemen*, published in Poland in 1959, though many of its stories were written shortly after the war when Borowski was in his early twenties. Dismissing Hochhuth's premise that we lack the imaginative faculties to be able to envision Auschwitz (formulated, ironically, years *after* Borowski's suicide in 1951), Borowski chronicles the divorce between reader expectation and inmate behavior through a casual and understated first-person narrative style. His narrators disarm us with the simplicity of their opening gambits, feigning a disinterest that is slowly undone by subsequent events. He refuses to supply us with guidelines for the inhuman tour we are about to begin, teasing our curiosity with hints of disorder that gradually invade our lingering innocence.

"All of us walked around naked," begins one of his stories, leaving his reader to wonder whether the speaker is Adam in the Garden of Eden or the resident of a nudist colony. The narrative then eases us into the place we call Auschwitz: "The delousing is finally over, and the striped suits are back from the tanks of Cyclone B solution, an efficient killer of lice in clothing . . ." The clinical clues are puzzling, but not yet sinister, until the sentence finishes: "and of men in gas chambers."[18] Generations from now, this passage may require footnoting; when Borowski wrote it, it needed only assent. What he forces us to assent to, however, violates every value that civilization presumes to cherish; Borowski's stories portray the systematic mutilation of such values in the Auschwitz he knew and experienced.

One test of a literature of Auschwitz is its candor in imitating the atmosphere of moral and physical mutilation that the Germans deliberately created in the camps. The notions of heroism and villainy so central to Frankl and Hochhuth in their visions of Auschwitz vanish from Borowski's literary horizon. Andrzej Wirth, a Polish commentator on Borowski's art, helps us to understand why; Borowski's scenario, he argues,

> has nothing to do with the classical conception based on the necessity of choice between two systems of value. The hero of Borowski's stories is a hero *deprived of all choice.* He finds himself in a situation without choice because every choice is base. The tragedy lies not in the necessity of choosing but in the impossibility of making a choice.[19]

When the goal of moral being is not virtue, but staying alive, then our sense of character loses its mooring in literature, Scripture, or philosophy, and succumbs to circumstance—the awful predicament that Borowski energizes in his Auschwitz stories.

In the culture of coping that defined existence in the deathcamp, the survivor depended for his life—at least for a brief time—on the death of someone else. If the tragic figure is one who through action or attitude rebels against his destiny, what are we to make of one of Borowski's narrators, who helps to drive victims from the cattle cars, unloads their belongings, watches them being led off to the gas chambers, feels rage at his involvement in their fate instead of pride at his mastery of his own, and finds in an attack of nausea little relief from an environment that dehumanizes everyone—murderer, victim, and survivor? Life gestures are contaminated by death or become death gestures themselves. The narrator spends his rage in impotent silence, bereft of meaningful choice: "The air is filled with ghastly cries, the earth trembles beneath me, I can feel sticky moisture on my eyelids. My throat is completely dry." Earlier he had broken his silence by explaining to some Greek inmates in what he called "crematorium Esperanto" the challenge that lay before the Canada Commando, in which they worked: *"Transport kommen, alle Krematorium, compris?"* In the global idiom of mass murder, words do not dignify and communication brings neither community nor communion. He is victim himself of what he calls the only permissible form of charity in Auschwitz, the camp law dictating that "people going to their death must be deceived to the very end."[20]

Among Borowski's most important contributions to the literature of Auschwitz are his portraits of what Levi would later name the "functionary-prisoners" in the camp, those squad leaders or Kapos or other inmates who through luck or manipulation joined the internal power hierarchy and thus entered the gray zone of its moral life, prolonging their own temporary survival amid the murder of others—before them, around them, often in their place. People staying alive, he suggests, must also be "self-deceived" about the origin of their survival, though the very consciousness of this fact lurks menacingly beneath the façade of indifference that usually dominates his narrator's voice. He is concerned with the state of mind bred by being among the "privileged," and the psychological price one pays to remain there.

But the internal power hierarchy in Auschwitz was neither exclusively male nor non-Jewish, though this is the prevailing rule in Borowski's fictional vision. Sara Nomberg-Przytyk's *Auschwitz: True Tales from a Grotesque Land,* in a series of interconnected stories, chronicles the odyssey of a Jewish woman who joins the ranks of female functionary prisoners before our eyes, learning through the process what one must discard of one's civilized moral and psychological baggage in order to gain and retain that status. Although she will be low down in the hierarchy, Nomberg-Przytyk's narrator is not blind to the gray zone she is soon to enter:

> The SS men saw the splendor in which the camp functionaries lived, but all this took place with their silent approval. It was a devilish system in which the SS men and the functionaries were united by a chain of cruelty. The contrast between their splendor and our misery kept them constantly aware of what they stood to lose in the event that they failed to carry out the orders of the SS men. They used whatever methods were necessary to assure their own survival and their relatively comfortable way of life. If the voice of conscience chanced to awake in them, they would quiet it continuously with the same arguments: "We suffered so much in the first few years. We lived through those hard times. Now we are not going to die for the sake of some dirty *zugang* [new arrival]."[21]

Such a passage may tempt us to believe that selfishness and brutality were the vital conditions for staying alive in this milieu, but the narrator discovers that motives in Auschwitz are far more baffling, controlled not by some inner system of values, but by circumstances unrelated to one's will.

She remembers thinking, for example,

> that in Auschwitz there was nothing more important than trying to help your fellow sufferers and yet, at the same time, how immoral it was to decide whose suffering should be alleviated and whose should continue unabated. Who had given us the right to condemn or to save another? In Auschwitz there was no fairness in the merciless struggle for survival. Those with scruples died isolated and abandoned. That was the new order of the concentration camp.[22]

This is what Wirth means when he suggests that in Borowski's Auschwitz all choices are base, including the decision to aid a fellow sufferer. If the literature of Auschwitz can help us to understand nothing more than this troubling but truthful paradox, it will have vindicated its vexing challenge to the imagination.

In October 1944, Nomberg-Przytyk's narrator is working as a clerk in the hospital barrack. One evening, while she and her comrades are discussing the difference between conscious and unconscious death, the other clerk in the infirmary interrupts with a story that freezes the momentum of their abstract debate. She tells of a group of 156 Polish girls from Krakow, who had been sent to the clinic, so they thought, for examination before being shipped to work in Germany. "They were talking loudly, laughing, never dreaming that they had been horribly deceived and that the *Leichenauto* [vehicle serving as a hearse] was coming for them in about an hour." Clearly they are "fellow sufferers" who, as the functionary prisoners know, are about to be killed by phenol injection. How does one define one's human role at a moment like this? "Perhaps I should shout it out to them," says the speaker, " 'Calm down! Don't laugh. You are living corpses, and in a few hours nothing will be left of you but ashes!' Then what? Then we attendants would go to the gas chambers and the women would die anyway."[23] Suddenly the question "Is it better for a human being to know that he is about to die?" assumes dimensions of complexity that no prior system of belief allows us to simplify.

Nomberg-Przytyk shares Borowski's talent for sketching the rupture between Auschwitz reality and conjectures about it—the prisoners' and our own. "We didn't tell them the truth," the speaker says,

> not out of fear for our own lives, but because we truly did not know what would be the least painful way for the young women to die. . . . If we told them what was in store for them, then a struggle for

> life would ensue. In their attempt to run from death they would find only loneliness, because their friends, seeking to preserve their own lives, would refuse to help them.[24]

Difficult as it is for us to imagine, not to say concede, mutual support becomes an academic question when the issue is death by phenol injection. Courage and truth itself shrivel into privileged virtues reserved for those living beyond the ominous shadows of the gas chamber.

In this instance, the young women themselves eventually discover what is in store for them, and a terrible outcry erupts. They are surrounded and beaten, and then each "was dragged screaming, by two SS men, into the presence of Mengele," who presumably administers the fatal injection. Some of the remaining victims try to run away. "Then the dogs were set on them. Their deaths were completely different from the deaths of the first batch of women who went to their deaths unknowing. Who knows which death was more difficult, but the first group seemed to die more peacefully." This story, called "The Verdict," suspends judgment but goads the reader into the role of juror through its conclusion: " 'I still don't know whether we should have told the women about the death that was waiting for them. What do you think?' None of us said anything."[25]

The reader is left speechless too, silenced by the sorrowful if contrary fates of victims *and* survivors. Is there such a notion as "complicity through the eyes," by which a witness is diminished simply because of what she has seen? When Nomberg-Przytyk's narrator is freed, she does not rejoice. "I felt comfortable, warm and clean," she admits. "But I was not happy. I did not know why. Again and again I repeated to myself the refrain: 'Be happy, you are free.' But this did not help. I was sad. Sadness strangled me."[26] Her response will seem odd only to those who have not immersed themselves in the literature of Auschwitz, where they are forced to discover how closely woven, morally and emotionally, are the stories of those whom Primo Levi, in a striking image, has called "the drowned and the saved."

If the saved remain tainted in their memories by the misfortune of the drowned, this is one of the melancholy bequests of the camp experience. "It was a logical consequence of the system," wrote Levi: "an inhuman regime spreads and extends its inhumanity in all directions, also and especially downward; unless it meets with resistance and exceptionally strong characters, it corrupts its victims and its opponents as well."[27] A corollary "logical consequence" of Auschwitz, to be extracted by the diligent moralist, is unfortunately this: there was no rein on the shame, humiliation, and torment that the Germans could inflict on their prey, no check to their malice, brutality, lust for ruin; their talent for atrocity was unlimited. Goodness, on the contrary, was curbed among the victims at every turn, by fear, hunger, thirst, confusion, illness, and despair. The *will* to compassion may have remained intact, but its power to oppose the ungoverned ferocity of the camps faltered before the sterile cruelty inherent in the system.

Few writers in the tradition of Auschwitz convey this painful truth with the dense immediacy of Charlotte Delbo, whose trilogy *Auschwitz et après (Auschwitz and After)* explores the fragmenting of the self and the uncoupling of its milieu that were the most enduring legacies of the camp experience.[28] For Delbo, Auschwitz was simply a place of unnatural, premature dying; her art represents a resolute search for a prose and poetry equal to this dismal fact. In her vision, the self is inseparable from the cold, hunger, and exhaustion that slowly erode its substance, until the crust of dignity formerly enclosing a human being loses its protective value and decays. She insists that we join her in witnessing what remains, as she and her fellow prisoners peer from a barrack window:

> At first we are not sure what we see. It is difficult at first to distinguish them from the snow. The yard is full of them. Naked. Lying close to each other on the snow. White, a white that looks bluish against the snow. Their heads are shaven, their pubic hairs are straight and stiff. The corpses are frozen. White with brown nails. Their upturned toes are truly ridiculous. Terrible, ridiculous.[29]

Is unaccommodated woman no more than this? Behind the logical consequences of Levi's Auschwitz, legitimate as they are, lies the physical assault on the body that mirrored its indelible nucleus.

Into the heart of this nucleus Delbo's incantatory prose lures us, groping toward images to match the abrupt reversals implicit in her theme:

> Standing, wrapped in a blanket, a child, a little boy. A tiny shaven head, a face in which the jaws and the brow ridge stand out. Barefoot, he jumps up and down without stopping, with a frenzied movement that makes one think of that of savages dancing. He wants to wave his arms too to keep warm. The blanket slips open. It is a woman. A skeleton of a woman. She is naked. One can see her ribs and hip bones. She pulls the blanket up on her shoulders and continues to dance. A mechanical dance. A dancing skeleton of a woman. Her feet are small, thin and bare in the snow. There are living, dancing skeletons.[30]

This is a true art of revelation, though not in the familiar sense. Unlike Frankl, Delbo has no qualms about exposing a culture built on mistaken identities, enticed by traditions of Enlightenment and Romanticism into forming idealized versions of the inviolable self. Auschwitz has disfigured those traditions, leading Delbo to focus on the violated human form and to ask how such fragmentation might be integrated into the ambitions of future generations.

The irony of such a quest, not to say question, does not escape Delbo herself. She follows her description of the living, dancing skeleton with an early example of what has become a postmodern fashion—a self-reflective admission of the artifice of art: "And now I am sitting in a café writing this story—for this is turning into a story."[31] Can a literature of Auschwitz *ever* span the chasm between what we were and what the camp's very existence has made us a part of? Throughout her narrative, Delbo pays tribute to the women friends who supported her at those moments when she felt unable to go on, as if one impulse of her story were to reaffirm the strength of human community despite the assault of Auschwitz on its spirit. But in the end, this support proves a wan comfort; she records the collapse of identity, the "failure" of character, the splintered unity, the merging of women with a mute, unfruitful earth.

The last vignette in *None of Us Will Return* (the first volume of *Auschwitz and After*) is called "Springtime," an ironic dirge to the season of renewal from whose solace Delbo and her friends seem endlessly barred:

> All these lumps of flesh which had lost the pinkness and the life of flesh were strewn about in the dusty dried mud, were completing the process of withering and decomposing in the sunlight. All this brownish, purplish, gray flesh blended in so well with the dusty soil that it required an effort to pick out the women there, to make out empty breasts amid this puckered skin that hung from women's chests.[32]

As Delbo the artist composes, her characters "decompose," and this is the challenge that a literature of Auschwitz will always have to face. The realities of the camp continue to contradict the premises of form and of language itself, resulting in a split that may in fact define the bond between the writer and this material, and our possible access to it.

In the second volume of her Auschwitz trilogy, *Une Connaissance inutile (A Useless Knowledge)*, Delbo offers a distilled variant on this dilemma:

I'm back from another world
to this world
that I didn't leave
and I don't know
which is real
tell me have I come back
from that other world?
As for me
I'm still there
and I'm dying
back there
every day a bit more
I die again
the death of all those who died
and I no longer know what's real
in this world
from the other world-back-there
now
I no longer know
when I'm dreaming
and when
I'm not dreaming.[33]

We have only to compare the returns of Odysseus and Aeneas from their "other world-back-there" with Delbo's to see how radically a demythologized literature of Auschwitz differs from traditional epic encounters with the realm of death. Odysseus carefully keeps the dead from profaning his living person, while Aeneas comes back from his visit to the underworld with a happy prophecy of a future civilization. Neither would understand Delbo's doom-laden line "I die again / the death of all those who died," a line that stretches the circle of its recruits to include its audience as well.

The experience of Auschwitz, like all of the Holocaust, cannot be left behind. Nor do we return from our encounter with its literature unblemished. Instead, like Delbo, Levi, and all the rest, we face the necessary burden of adjustment. When Delbo admits "I return / from beyond knowledge / now I need to unlearn / otherwise it's clear / I couldn't go on living," she intends not to slight her past, but to invite us to share with her the twin vision that a journey through Auschwitz has etched on our culture. We pay a price for learning how to imagine what happened; then we add to our debt by feigning that beyond those mounds of corpses and heaps of ashes a chaste future is still feasible: "because it would be too stupid / in the end," as Delbo agrees, "for so many to have died / and for you to live / without making something of your life." But she frames this with a more somber paradox, one that echoes hollowly through the Holocaust universe, leaving us little but a bleak query to kindle hope:

> I've spoken with death
> so
> I know
> how useless were so many things we learned
> but I gained this knowledge at the price of suffering
> so great
> I wonder
> if it was worth it.[34]

1991

9

Kafka as Holocaust Prophet: A Dissenting View

omeone must have been spreading rumors about Franz Kafka, for without having done anything wrong, he was proclaimed one fine morning the prophet of the Holocaust. But when rumor congeals into widely accepted critical truth, the time may have arrived for an inquiry into its genesis and the sufficiency of its allegations. "Naturally," Kafka wrote at the very end of his *Letter to His Father*, "things cannot in reality fit together [*aneinanderpassen*] the way they do in my letter; life is more than a game of solitaire [*Geduldspiel*]"[1] The temptation to read—or misread—Kafka's imaginative world as an anticipation of the death-camp universe has led to a similar confusion between art and life, though, as Kafka admits in the conclusion to that letter, with certain modifications one may reach approximate truths concerning the connection between what is written and what happens "in reality." Kafka cautiously limited his reference to the bond between his father and him. We tread on more hazardous ground when we extend the margins of this reference to the link between Kafka's art and the catastrophe of European Jewry.

Others have been less circumspect than Kafka, and their enthusiasms for his prescience have established an entrenched view that has become one of the unexamined shibboleths of Kafka criticism. What, for example, are we to make of George Steiner's dramatic insistence that "Kafka heard the name Buchenwald in the word birchwood" (Kafka, incidentally, had he heard it at all, would have heard it in "beechwood," not "birchwood")? Or his even more elaborate pronouncement that Kafka "prophesied the actual forms of the disaster of Western humanism"? Assuming that this disaster culminated in the gas chambers and crematoria of central Europe, one searches through Kafka's pages in vain for emblems, images, or metaphors of its "forms": ashes, chimneys, boxcars, mass graves, seared flesh, twisted corpses, physical annihilation. Excluded from what Steiner and many others call Kafka's "nightmare-vision"[2] are balancing features of his art, like parody, irony, humor, and the constant yearning for reconciliation that kept his vision from tumbling into an abyss of utter despair. Reinterpreting Kafka in the light of later experience is a perfectly legitimate response of modern tradition to his individual talent. But appropriating Kafka to confirm a tradition that may not exist at all is another matter entirely, which risks distorting Kafka's achievement and obscuring the challenge facing the Holocaust writer in search of his own authentic inspiration.

Steiner's view of Kafka's "fearful premonition" is based on dubious evidence. *The Trial,* he asserts,

> exhibits the classic model of the terror state. It prefigures the furtive sadism, the hysteria which totalitarianism insinuates into private and sexual life, the faceless boredom of the killers. Since Kafka wrote, the night knock has come on innumerable doors, and the name of those dragged off to die "like a dog" is legion.[3]

Charitable opinion would call this a careless reading of the novel, though generations of students have been beguiled by statements like these to see *The Trial* as an uncanny prevision of Gestapo tactics. But the two men who come for Josef K. at the end (like the warders who arrest him in the first chapter) are neither furtive nor sadistic. K. is not frightened when he leaves, nor are his executioners brutal; indeed, terror is a mood distinguished by its absence from the novel. There is in fact no knock on the door, and K. is not "dragged" off at all. The "gentlemen" who come for Josef K. behave with firm but ceremonious courtesy; the "faceless boredom of the killers" is imposed on Kafka by the critic's fiat, not critical insight. The final noun of the novel, as we have it, is neither "horror" nor "terror," but "shame," which shifts the focus of Josef K.'s manner of dying, "like a dog," away from his executioners and back into his own consciousness, which has been the novel's main terrain from the beginning.

If Steiner's serious (and valuable) explorations of the question of language and writing in a post-Holocaust world have led him to exaggerate Kafka's oracular role, we should be less surprised to find other distinguished commentators lapsing into similarly excessive claims. Curiously, the idiom changes according to the values of the critic; Kafka is not interpreted, but co-opted, with a formulaic assurance that dispenses with the need for substantive evidence. Bertolt Brecht's voice is almost predictable:

> We find in [Kafka] strange disguises prefiguring many things that were, at the time when his books appeared, plain to very few people. The fascist dictatorship was, so to speak, in the very bones of the bourgeois democracies, and Kafka described with wonderful imagina-

> tive power the future concentration camps, the future instability of the law, the future absolutism of the state *apparat,* the paralyzed, inadequately motivated, floundering lives of the many individual people; everything appeared as in a nightmare and with the confusion and inadequacy of nightmare.[4]

Of course, the practical ideologist in Brecht could hardly find Kafka's concern with the metaphorical ambiguities of the law and state exemplary art. His remarks nonetheless reinforce the "exemplary" Pavlovian equivalence which has developed over the years between "Kafka," "nightmare," and "concentration camps." The only item missing is the original stimulus which inspired the association, as if a conspiracy of agreement existed among "knowledgeable" Kafka readers whose shared premises required no confirmation.

Holocaust analogies infiltrate the imaginations of some of Kafka's most original critics, suggesting perhaps a latent need to believe that only future history could validate an eccentric literary vision that past and contemporary events were insufficient to explain. How else can we justify Günther Anders's gratuitous observation that "K.'s living room in the gymnasium in *The Castle* is hardly more fantastic than the parlours [in the commandants' homes] in the extermination camp"? The incongruous juxtaposition of the normal with the bizarre was hardly discovered by the Nazi worldview, yet readers persist in drawing Kafka within its perimeters. Written so soon after the Second World War, Anders's *Kafka: Pro et Contra* inevitably is suffused with memories of the recent catastrophe. But the proximity in time has led to analogical formulations that do a disservice to Kafka's intentions and effects. "At the beginning of the story *In the Penal Colony,*" says Anders, in a now-familiar comparison, "an officer shows an explorer a highly complex and diabolically conceived execution-machine, such as the world had never seen before the time of Hitler's instruments of mass murder."[5] Since the unexamined (implicit) simile is not worth preserving, let us examine this one. Not much reflection is required to realize that Hitler's methods of extermination have no more provenance in Kafka's "peculiar apparatus" (*eigentümlicher Apparat*) than Kafka's execution-machine finds its large-scale fulfillment in Hitler's deathcamps. Simple rifles and machine guns were the first "instruments" of mass execution used by Hitler's SS; but even death by some form of gassing and disposal of bodies by cremation were neither original nor diabolical as forms of killing, though the impulse to employ them for genocide certainly was. Indeed, the whole point of Kafka's machine is that it is not an instrument of mass murder, but designed to impose on a particular victim the sentence appropriate to his particular "crime."

The genesis of Kafka's "peculiar apparatus" requires no ex post facto confirmation by the Satanism of Hitler and the Holocaust. Late-nineteenth- and early-twentieth-century culture's response to the potential Satanism of the machine furnishes sufficient background. Consider the following statement: "The machine process pervades the modern life and dominates it in a mechanical sense. Its dominance is seen in the enforcement of precise mechanical measurements and adjustments and the reduction of all manner of things, purposes and acts, necessities, conveniences, and amenities of life to standard units." Although it is unlikely that Kafka ever read the work from which these lines are taken—Thorstein Veblen's *The Theory of Business Enterprise* (1904), written when Kafka was just about to embark on a literary career—it reflects one attitude current in the West toward the nineteenth-century glorification of technology. Kafka's story, among other things, parodies that excessive reverence for this technology. Veblen's comments on how process subtly intrudes on existence form an incidental gloss on the victimization of Josef K. as well:

> Mechanically speaking, the machine is not [man's] to do with it as his fancy may suggest. His place is to take thought of the machine and its work in terms given him by the process that is going forward. . . . If he fails of the precise measure, by more or less, the exigencies of the process check the aberration and drive home the absolute need of conformity.

Thus argues the Officer to the Traveler in "In the Penal Colony," too; but when the process breaks down, and Kafka's "apparatus" collapses, the system goes awry and unpredictable (and inexplicable) catastrophe ensues. As the discipline of the machine invaded more and more of modern culture, Veblen apprehended, the result—unless some remedy were found—would be "an ever-weakening sense of conviction, allegiance, or piety toward the received institutions."[6] Veblen, in other words, quite unconsciously predicts the fate of the old order in Kafka's Penal Colony, as Kafka's less didactic and more somber vision evokes the disarray resulting when the attempt to "clarify" native values (*erklären* is one of the key words of Kafka's story) is met by a stubborn failure or refusal of the outsider to "grasp" (*begreifen* is another) the meaning of that peculiar apparatus for the rise and fall of a civilization. The Officer's explanation (*Erklärung*), in other words, is inconceivable (*unbegreiflich*) to the Traveler, and the reader feels alienated by the absence of dialogue between the two points of view. And this is authentic Kafkan terrain.

Kafka's "In the Penal Colony" ratifies the attitude of many of his contemporaries that the machine would eventually consume the individual arrogant enough to believe that he permanently controlled its processes. Some German Expressionist drama and film—Fritz Lang's *Metropolis* comes to mind—illustrate the dangers of such an illusion; and even Charlie Chaplin's later *Modern Times,* in a comic vein, verifies this thematic concern. As for Kafka's anticipating Hitler's instruments of mass murder in "In the Penal Colony," this seems now to be a superfluous, not to say irrelevant, observation. Günther Anders's insistence that "Kafka's 'marionettes' are prophetically enacting through their grotesque selves and experiences the most ghastly occurrences of our time," rather than of their own, betrays an urgent but misdirected modern need to find in past art (if not past history) "logical" precedents for the unprecedented illogic of the Holocaust. Anders's conclusion that "death in concentration camps has been the fate of those to whom society could not or would not assign a particular function"[7] blurs the distance between Kafka's vision and the Holocaust by ignoring the fact that Hitler's victims (unlike some of Kafka's characters) had specific functions in society which they were perfectly content with, but which their murderers refused to acknowledge.

Theodor Adorno comments on the tendency of some critics to assimilate Kafka "into an established trend of thought while little attention is paid to those aspects of his work which resist such assimilation and which, precisely for this reason, require interpretation." This is exactly the distinction which needs to be made when considering Kafka's alleged ancestral relationship to Holocaust literature, yet even Adorno occasionally fails to heed his own cautionary advice. Kafka does not "directly outline the image of the society to come," Adorno concedes, "—for in his as in all great art, asceticism toward the future prevails—but rather depicts it as a montage composed of waste-products which the new order, in the process of forming itself, extracts from the perishing present." The question, however, is whether that "new order" drew on the waste products of Kafka's perishing present, or manufactured fresh resources of atrocity from raw materials bearing only tenuous resemblances to the refuse of Kafka's reality. Adorno may be correct in affirming that Kafka's epic course was the "flight through man and beyond into the non-human,"[8] but a sharp difference exists between the nonhuman of Kafka's vision and the inhuman of the Holocaust universe, which reflected not only the collapse and impotence of subjective consciousness but the disintegration—or "demolition," to use Adorno's more explosive metaphor—of the physical self. The nonhuman asserts itself in Kafka through the oppressive if uncertain presence of the Castle in the distance, the Officer's execution machine of "In the Penal Colony," even the tunnels and corridors of "The Burrow," which serve as refuge or trap. Some form of energy reverberates from these nonhuman sources of oppression and threat against which the victimized creature may measure his failing vitality and sometimes make a gesture of protest. A human Klamm with defined voice and shape may be an elusive possibility throughout the pages of *The Castle*, but Klamm the Official, a nonhuman authoritative presence dominating K.'s expectations and determining his strategy, bears little resemblance to the concrete inhuman threats to existence that crowd the pages of Holocaust literature. There, "confined space" is translated into the sealed boxcar, the gas chamber, the burial pit, the crematorium, replacing the interminable small rooms and offices of Kafka's world; even as breathing, afflicted by the odor of charred flesh and Cyklon B, replaces the stifling airlessness that makes Josef K. temporarily feel faint. The flight through man and beyond into the inhuman, with its own universe of appropriate imagery, separates Kafka's emphasis from that of his Holocaust successors.

Adorno's proposition that the "social origin of the individual ultimately reveals itself as the power to annihilate him" is a two-edged sword whose relevance to our subject depends on which edge we choose for our execution—or suicide. But it also depends on how we interpret "annihilation," and I suggest that it has associations for us (and for Adorno) that were undreamed of in Kafka's philosophy. Looking forward from the Holocaust, we may more easily understand (though not justify) the episodes of limited genocide that have spread across our landscape of violence. But looking backward from that same event, we find little continuity between Kafka's spiritually anonymous, misplaced, perhaps unplaced individual and the eliminated ones of our own time. Kafka's protagonists suffer from their own loss of spiritual continuity, and Adorno's conclusion that their lives represent trial runs for models of dehumanization would be irreproachable, if he had practiced the same asceticism toward the future which he found so admirable in Kafka. "The crucial moment," he argued, "towards which everything in Kafka is directed, is that in which men become aware that they are not themselves—that they themselves are things." But Adorno was not content to see this as a culmination of preceding impulses in nineteenth-century culture. Kafka would have had to search no further than Gogol—*The Overcoat* and *Dead Souls*, for example—to discover antecedents for the idea that men lose a dimension of their humanity when they find an equivalence of their selves in things. Moreover, the curious reader could find in Dostoevsky's Underground Man, and especially in Dmitri Karamazov, the sense of personal shame, humiliation, and inadequacy that overwhelms the individual with an impression of his own "bugness." Insect imagery provides one of the main strands of *The Brothers Karamazov*. What, then, but the irrepressible thrust of retrospective vision, combined with a stubborn resolve to see the Holocaust as a *necessary* expression of a cultural impulse that had been quivering for decades, could have driven Adorno to insist that "In the Penal Colony" and "The Metamorphosis" were "reports which had to await those of Bettelheim, Kogon and Rousset for their equals"? Specialists might find this an odd troika, since their accounts of the camp experience differ so widely in content and emphasis; Bettelheim, indeed, studiously avoided in his early essay "Individual and Mass Behavior in Extreme Situations" and the later *Informed Heart* the very emphasis that Adorno praised in Kafka, his "capacity to stand up to the worst by making it into language."[9] Moreover, equating Kafka's imaginative vision with survivor accounts of actual events, whether muted or gruesome, raises more questions than Adorno settles by his analogy.

Can it be that some unarticulated quality in certain émigré temperaments inspired them to search for such connections? Adorno mentions favorably Klaus Mann's contention that there was "a similarity between Kafka's world and that of the Third Reich," and though he knows that any direct political allusion would have violated the hermetic spirit of Kafka's approach, he still maintains that "it is National Socialism far more than the hidden dominion of God that his work cites." But of course these are not the only options, nor need they be mutually exclusive, should we choose to be so specific in our interpretations. Kafka blurs traditional distinctions between village and castle, thus spreading ambiguity and obscurity in both directions, not merely into the realm of the Other. Adorno transforms the behavior of Castle officials into a kind of proto-fascism, though its effects are never sinister or violent, simply confusing; nor is it ever clear what is more responsible for the unhappiness of villagers like Barnabas and his family—Castle officials' misuse of power, or the characters' own excessively submissive consciousness. Surely Adorno intrudes his own political-economic convictions by writing that "Kafka's method was verified when the obsolete liberal traits that he surveyed, stemming from the anarchy of commodity production, changed into the forms of fascist organization." And when he adds that "it was not only Kafka's prophecy of terror and torture that was fulfilled," but that in *The Castle* "the officials wear a special uniform, as the SS did,"[10] then the pseudo-equivalence between Kafka and Nazism verges on the absurd. Castle officials, whom we virtually never see, whose behavior is reported more often than it is observed, and then usually on the basis of supposition or rumor, resemble SS men in their appearance and conduct no more than Titorelli the painter resembles Pablo Picasso.

Moreover, the number of officials who wore "special" uniforms in the Austro-Hungarian Empire was probably legion.

No one disputes that the spirit of terror and the practice of torture dominated life in Europe during the era of the Third Reich. The justification, if not the explanation, is to be found in the ruthless and brutal principles of Nazi ideology. Those who lived under its sway would have considered the frustrating, often dreamlike atmospheres of *The Trial* and *The Castle* paradise on earth. When a semblance of torture does appear in Kafka, as in the whipping episode in *The Trial*, the circumstances limit the intensity and the scope of the ordeal. The antics of the prisoner in "In the Penal Colony," as the Officer explains to the Traveler the details of the impending execution, introduce comic overtones to the scene, deflecting the reader's attention from what otherwise might have been judged a brutal procedure. Language, in Kafka, overlays the action like an anesthetic, allaying suspicion and inhibiting pure emotional response. This *was* an important if unintended legacy of Kafka to Holocaust literature, but in forging links between Kafka and the Holocaust, Adorno uses language for opposite purposes, to bludgeon the reader into accepting Kafka's prophetic role, even at the price of misrepresentation.

The strategy is worth considering in detail, since by reversing the sequence of historical reference, applying truisms of the Third Reich to unverified incidents in Kafka, Adorno replaces inference with identity, substituting established fact for the question of influence. "Arrest is assault, judgment, violence," he affirms, as if no difference existed any longer between Gestapo tactics and Kafka's indefinite world of guilt, shame, and accusation. Then he adds, and here we see the reversed sequence of historical reference at work: "The Party always allowed its potential victims a dubious, corrupt chance to bargain and negotiate, as do Kafka's inaccessible functionaries."[11] The temptation to rewrite Kafka's art to accommodate itself to Nazi reality is nowhere, perhaps, more lamentably illustrated than here. For what could Adorno have been thinking of? It is never clear, first of all, whether Kafka's "inaccessible functionaries" allow *anyone* to "bargain and negotiate": Josef K. and K. take up their own causes without any authorization from above, while Klamm's complimentary letter to K. in praise of the land-surveying work he has never done suggests that the Castle officials are ignorant of or perversely indifferent to the real situations of their potential "victims." These seek out intermediaries and send or await messages through their own volition, though their efforts are never acclaimed or even confirmed, and we are left suspecting that their attempts to establish contact with the functionaries may be misguided or self-deluded. As for a Nazi party allowing its potential victims (for which we must read Jews) a chance to bargain or negotiate, dubious and corrupt or not—fantasy replaces fact in this singular desire to confirm similarity where none exists. But Adorno's assertion illuminates the methodology by which Kafka has earned the reputation of prophet of the Holocaust. At this point, it is hardly necessary to remind ourselves how much of that reputation originates in Kafka's commentators, not in Kafka himself.

By now it should be evident that the need to make Kafka a progenitor of the Holocaust universe involves much more than the accident or miracle of artistic intuition. Establishing precedents for the unprecedented allays the puzzled conscience of a dismayed generation that still has trouble living with the unaccountability of the history of its own time. By what perverse imagining, other than wishing to invent a nonexistent precedent, could Adorno have made Gisa, the "blonde schoolmistress" of *The Castle,* stem from "the pre-adamite race of Hitler *Jungfrauen* who hated the Jews long before there were any"? Or contrast Amalia's refusal of Sortini with "the rite of the Third Reich" according to which "girls were not permitted to refuse [their sexual favors to] medal-of-honour winners"?[12] No Castle rule exists requiring Amalia to submit, nor does Sortini ever appear to confirm his alleged demand. The reduction of Amalia's family to social pariahs, attributed by Adorno to "tribal rite," might more exactly be described in the context of Kafka's novel as a result of community prejudice, reinforced by Amalia's family's self-imposed and exaggerated guilt and humiliation. Such conclusions expose Adorno, not Kafka, to the charge of seeing Buchenwald in the term "birchwood," though not at all in the sense that George Steiner intended.

Striking parallels certainly exist between Kafka's world and the world of Holocaust literature. But the alienated consciousness, Kafka's persistent theme, becomes "physicalized," as it were, in the later period, where the problem is no longer a matter of finding countermeasures to sustain perception, but the sheer necessity of maintaining physical existence. The threatened annihilation of the body breeds different challenges than the threat to the dignity of integrated consciousness, both for the victim and for the literary imagination. In a universe where man, as Elie Wiesel has written, was little more than a starving stomach, the leisure of consciousness was a privilege granted to only a fortunate few. Writing in June 1938, when consciousness could still function healthily (if precariously), Walter Benjamin struggled with some of these distinctions, ultimately proving to be much more of a prophet than the Kafka he was discussing. Looking backward (and simultaneously into the very near future), Benjamin called Kafka's world "the exact complement of his era which is preparing to do away with the inhabitants of this planet on a considerable scale." It is possible that Benjamin was referring to the "technology of modern warfare," which he had mentioned just a few lines earlier; but his subsequent observation raises the prospect of a more sinister meaning: "The experience which corresponds to that of Kafka, the private individual, will probably not become accessible to the masses until such time as they are being done away with."[13] Is this a melancholy concession to the irresistible juggernaut of history, embodied in the imposing power of the Third Reich, an epitaph to the impotence of its victims, or simply his own laconic foreknowledge of an inevitable doom?

Himself dwelling in an intermediate realm between private vision of individual fate (Kafka's theme) and public recognition of the prospects of mass destruction, Benjamin wrestled with the temptation to make Kafka the precursor of his generation's generalized destiny (1938 was too early to be more precise), but he resisted its appeal. Kafka's experience, he concluded, was "based solely on the tradition" to which he surrendered: "[T]here was no far-sightedness or 'prophetic vision.' " Kafka was a listener, not a seer, and what he "heard" was not the specific forms of the disaster of Western humanism, but the failure, or at least the illness, of the past, which had seriously infected his present. The devastation this disease might wreak as it metastasized in the future into racist ideology and policies of extermination were not even foreseen in detail by Benjamin, himself one of its early victims. His own formulation of Kafka's position vis-à-vis a present that was soon to consume him challenges us with a difficult insight into the essential distinction between the dilemma of Kafka and the ordeal of Benjamin and his fellow Jews: "If one says that [Kafka] perceived what was to come without perceiving what exists in the present, one should add that he perceived it essentially as an individual affected by it. His gestures of terror are given scope by the marvelous margin which the catastrophe will not grant us."[14] Kafka was a haunted writer, but not a hunted one. When Benjamin became a hunted writer too, he lost, or convinced himself that he had lost, the "marvelous margin" that enabled Kafka to confront life-threatening gestures of terror with life-preserving gestures of the imagination. The limits of Kafka's "prophetic vision" are bleakly illuminated by Benjamin's suicide. The frustration of consciousness instigated by Law Court judges and Castle officials neither resembles nor prefigures the futility and despair that closed in on Benjamin at the end. Like Kafka's sisters, also victims of the Third Reich, he must have discovered what Kafka himself never knew: the utter impotence of consciousness when the enemy is not a Ruthless Power, but "merely" a ruthless power, with no illusions (however tentative) joining it to a transcendent world of Higher Authority.

Whether we speak of Josef K.'s arrest, the Officer's apparatus in "In the Penal Colony," K.'s search for confirmation in *The Castle,* or the fate of the Jews in the Third Reich transformed into literary vision, the writer intent on mythifying, symbolizing, or simply *imagining* the meaning of victimization in our time must ask himself, as Anthony Thorlby does of Kafka, whether "there is a necessary connection between any fact of existence—be it an object, event, or experience—and the language used to express it." One might assert, as Thorlby does (though certainly his position is arguable), that we violate the spirit of Kafka's explorations into the potential of language when we "try to translate his stories back into the circumstances from which they may have sprung in his life, or into what they could theoretically indicate in the life of our time."[15] Certainly the following diary entry from July 1913 must give pause to anyone seeking the sole origin of the machine in "In the Penal Colony" in some historical or technological detail, past, present, or future:

> This block and tackle of the inner being. A small lever is somewhere secretly released, one is hardly aware of it at first, and at once the whole apparatus is in motion. Subject to an incomprehensible power, as the watch seems subject to time, it creaks here and there, and all the chains clank down their prescribed path one after the other.[16]

Perhaps we would be less prone to attribute foresight to Kafka, and find anticipations of Hitler's torture instruments in the Officer's machine, were we to pay more attention to lively imaginings like these, which nurtured the central event of "In the Penal Colony" without referring explicitly or implicitly to external circumstance at all.

Kafka lived and wrote from what he called an "inner plight,"[17] a constant tension that fed his imaginative universe. The world of Holocaust literature is sustained by a vision of victimization that is the antithesis of Kafka's, an outer plight which could not be mitigated by the qualities of being so vital to Kafka's art: age, sex, health, family situation, talent, temperament, and a certain kind of solitude. By externalizing his inner plight, Kafka could transform it into what Thorlby calls "essentially mythic material, the sheer fact and flow of existence, which has no necessary connection with [Kafka's narrators'] own standards and laws."[18] The Holocaust writer, who unlike Kafka must convey essentially *historical* material, enjoys no such freedom. Kafka's impulse is to mythicize the alienating universe where his animal and human narrating consciousnesses dwell, whereas the challenge to the Holocaust writer is to avoid this temptation, since the concrete images on which he builds his literary vision are more powerful (and less distracting) than any metaphors of existence his mythifying impulse might conjure up. The violation of self and the indifference to self that Josef K. and K. must contend with are connected to an ambiguous Other which breeds defensive truculence in its victims by its very vagueness; the characters in Holocaust fiction may delude themselves about the dangers before them, but there is never anything vague or unspecific about the forms or agents that threaten them. The "enemy" is only too vivid.

When Jorge Semprun's narrator (like Semprun, a survivor of Buchenwald) in *The Long Voyage* summons up a "special symbol" of the deathcamp universe (which afflicts him with the inner plight of inescapable memories), he invents no metaphors, no mythical monuments looming from the earth like avatars of some ancient civilization, no Great Wall of China or Tower of Babel to be reshaped into reminders of modern man's alienation. His images are direct and unambiguous: his "castle" is *there;* his law courts are accessible to visual memory:

> years later, remaining standing the longest, like the remembrance, or rather the evidence, the special symbol of that whole, the massive square chimney of the crematorium, till the day when the roots and the brambles shall also overcome that tenacious resistance of brick and stone, that obstinate resistance of death rising among the waves of green covering over what was an extermination camp, and those shadows of dense black smoke, shot through with yellow, that perhaps still lingers over this countryside, when all the survivors, all of us, have long since disappeared. . . .[19]

Such symbolization forestalls, even prohibits, the kind of internalizing of experience that animates narrative consciousness in Kafka. Semprun's survivor-narrator acknowledges images rising from the soil of history, not myth, images which he can share with his reader's consciousness, insofar as the reader submits to their promptings. But they are literal promptings, not literary ones. Kafka's Castle and Law Courts, his gigantic vermin and execution-machine, also prompt the reader, but with associations originating in the artist's and the sympathetic reader's inner plight, not the history of their time.

Just as the narrator's consciousness in *The Long Voyage* is dominated by a single crucial moment, toward and from which all memory and anticipation are directed—arrival at the concentration camp—so all Holocaust literature draws on a single event, whether represented in a particular work or not: extermination in the deathcamp universe.

But Kafka resists alluding to recognizable events, which makes the reader's immersion in his world totally different from the reader's involvement in the world of Holocaust literature. For example, Kafka seems to have made a calculated effort to insulate himself from the distractions of the First World War. His diaries and notebooks contain a handful of entries on that event, as if the mass-dying of those years had never penetrated his consciousness. He records the outbreak of war with disarming nonchalance: "August 2. Germany has declared war on Russia. —Swimming in the afternoon."[20] And on Armistice Day, November 11, 1918, he dispatches three communications to his publisher Kurt Wolff, including a postcard announcing that "I am sending the manuscript of the 'Penal Colony' by special delivery, registered mail along with a letter,"[21] reminding us only how long he had delayed the publication of this work, completed four years earlier. Kafka did not ignore the war, but regarded it as an intrusion rather than an inspiration, as he regarded everything which distracted him from his mission as a writer. "The thoughts provoked in me by the war," he wrote in his diary, "resemble my old worries over F[elice Bauer] in the tormenting way in which they devour me from every direction. I can't endure worry. . . ."[22] Those who extract anticipations of the Holocaust from Kafka's vision not only misapply the content of that vision, but also ignore the way in which he manipulated the items of external reality for his special solipsistic ends. The genesis of his imaginative world from a scrupulously observed "inner plight" separates and, one might even say, alienates him from the practitioners of Holocaust art who are bound to external reality by the very nature of the event they seek to represent.

Kafka's imagination worked in an entirely different way. The primary source of his artistic inspiration is nowhere more dazzlingly illustrated than in the opening line of a letter to Max Brod of July 1922: "Dearest Max, I have been dashing about or sitting as petrified as a desperate animal in his burrow. Enemies everywhere."[23] The following year, he wrote "The Burrow." Kafka transmuted the noise of children playing outside his window and the sounds of hammering and loading of timber from the nearby railroad station into the uncertain and even unverifiable sources of fear and anxiety that so distress the creature of his story, and make its existence unbearable. Perhaps it exaggerates the dangers that threaten, as Josef K. underestimates the gravity of his situation by depending too much on lawyers and legal proceedings in responding to his "case." But even if Josef K. had followed the prison chaplain's advice and taken his case more seriously as a comment on the human condition, there is no evidence that he would have been able to identify more clearly the source of the accusation against him. Kafka's victims inherited unverifiable anxiety and unspecified guilt as the natural terrain of their being.

Superficially, it would seem that we have finally established a crucial link between the worlds of Kafka and the Holocaust, since surely the victims of that later universe inhabit the same terrain. But Josef K. and K., like the creature in "The Burrow," are never sure that they are dealing with a substantial reality, an authority literally responsible for their ordeal. In the case of the victims of the Holocaust universe, on the other hand, whether the threat to existence appears in the foreground, as in the stories of Tadeusz Borowski and the memoir-visions of Charlotte Delbo, or indirectly and suggestively, as in the short narratives of Jakov Lind and the fiction of Aharon Appelfeld, the reader cannot escape the ultimate burden of physical atrocity: death by injection, exhaustion, beating, disease, shooting, gassing. The catastrophe of men destroyed replaces the dilemma of man dying. The difference between the two imaginative realms is vividly defined by Josef K.'s naive legalistic gesture following his arrest—"Here are my papers, now show me yours, and first of all your warrant for arresting me"[24]—and the opening line of Jakov Lind's "Soul of Wood," whose primary setting is a lunatic asylum populated by doomed Jews: "Those who had no papers entitling them to live lined up to die."[25] In Kafka's fiction, the destiny of "those" is not a looming concern; it dramatizes instead the fate of "me," however painful, whether the execution of Josef K. or the expiration of Gregor Samsa. Just as, in Jean Améry's striking formulation, "no bridge led from death in Auschwitz to *Death in Venice*,"[26] so no bridge leads from the death of Josef K. to mass extermination. No matter how bizarre the circumstances of his impending death, he may still reflect in the penultimate moment of his life: "Were there arguments in his favor that had been overlooked? Of course there must be. Logic is doubtless unshakable, but it cannot withstand a man who wants to go on living."[27] Invalid as these conjectures turn out to be, they still reflect his confidence in verbal argument, the invaluable resources of language to convince "someone" that this life is worth rescuing.

A similar confidence exists for those who seek to make literature into a form of prophecy, and Kafka a precursor of the Holocaust, by rescuing from the irreducible irrationality that led from Jewishness to the gas chamber a trace of logic, at least in imaginative foresight. If language preserves a semblance of order that disintegrates in reality, if after the fact we really can see Buchenwald and Birkenau in the groves of Kafka's dark woods, then artistic vision, however obscurely, affirms continuity and admits some light into the absolute blackness of physical annihilation. It is a consolation devoutly to be wished. But it reflects our own spiritual strategies, rather than Kafka's. Martin Walser argues convincingly that Kafka "has already completed the task of transmuting reality before he comes to write his fiction,"[28] and if this is so, then the process of art transmuting reality in Kafka is the very reverse of reality transmuting art in Holocaust literature. Remote officialdom may impose anxiety on Kafka's victims and generate a campaign for acknowledgment and just treatment; such options are excluded from Holocaust literature by the nature of the event. When Gregor Samsa's sister exclaims, "I won't utter my brother's name in the presence of this creature,"[29] Gregor's waning human identity receives a fatal blow and he soon dies. Although both Kafka and the Nazis favored the term, the *Ungeziefer,* or "vermin," of Kafka's story resonates in the fate of a single human being, whose own sense of unworthiness contributes to its potency as a weapon against him. But in Nazi usage "vermin" applied to the broad idea of *das lebensunwerte Leben,* lives unworthy of being lived, an idea no more implicit in Kafka's restricted image than was genocide in the "peculiar apparatus" of "In the Penal Colony." Indeed, one possible source of the heinous notion that victims in the Holocaust universe collaborated in their own extermination may be the unmotivated guilt and intuitive self-doubt that inspire the behavior of so many of Kafka's victims.

Striking parallels do not a prophet make. What the Holocaust writer (and reader) learn from the Kafka legacy is the incompleteness of Kafka's vision for a world ruled by the inhumanity of genocide. The parallel, which has led so many to turn it into a prophecy, is clear from a diary entry of February 1922: "Looked at with a primitive eye, the real, incontestable truth, a truth marred by no external circumstance (martyrdom, sacrifice of oneself for the sake of another), is only physical pain." The Holocaust survivor or victim, deprived of the consolations of traditional martyrdom by the anonymity of the killing methods and of self-sacrifice for others by the total power of the persecutors, would have recognized in Kafka's statement the features of his or her own ordeal. Moreover, just as Kafka's readers wrestled (and still do) with the implications of a vision that dispenses with the myth of redemptive dying, so readers (and writers) of Holocaust literature struggle with the meaning of that loss. But though Kafka demoted the myth, he did not abandon the vocabulary and imagery of mythology; in subsequent sentences he speaks of the "god of pain," and goes on to individualize and personalize the challenge of dying: "for the tubercular the god of suffocation." The Holocaust writer, confronted with the spectacle of mass atrocity rather than private dying, finds irrelevant and irreverent the idea of a god of suffocation, whose approach grows more bearable, in Kafka's words, if one can "partake of him in advance of the terrible union."[30] Kafka's "god of pain" represents one of the many *Gegner,* or opponents, who inhabit his universe: Law Court or Castle officials, Gregor's family, Georg Bendemann's father, the threatening presence in "The Burrow." Whether they exist in reality or in the victim's imagination, they humanize and sometimes dignify his ordeal by making opposition possible, if not often successful.

Precisely here the continuity between Kafka's imaginative universe and the world of the Holocaust ruptures, and the breach is really irreparable. Neither the Holocaust writer nor the Holocaust victim can invoke a god of suffocation or imagine some preliminary confrontation to soften the doom that is about to consume so many millions. Kafka's universe is predicated (like Camus's after him) on a stubborn, persistent, if ultimately futile instinct to struggle, however uncertain the enemy's configuration. But in the Holocaust universe, the source of stubbornness and persistence lies rarely in the abused victim but more often in the enemy, whose features are unmistakable and whose instrument of destruction, suffocation, has nothing to do with man's private fate or the vocabulary of myth that Kafka associates with it. Kafka's victims, like his readers, face a world full of lonely lives and lonely deaths, shrouded by spiritual and intellectual confusion and misunderstanding. It is a painful heritage, but with courage and the help of Kafka's art, which was adequate to its task, we can confront it. At this point in time, Holocaust literature must face a violation, not an expression, of man's fate in the modern era; vision pursues valiantly but in vain a vocabulary sufficient to the doom of mass extermination. Kafka never knew how fortunate he was to have suffered such a manageable challenge.

1986

10

Aharon Appelfeld and the Language of Sinister Silence

haron Appelfeld's art takes us on a journey into the realm of the unsaid; but it rejects the corollary idea, so often maintained by commentators on Holocaust literature, that the unsaid is necessarily unsayable. This distinction is at the heart of his imaginative vision. Much of his fiction invites us to experience not catastrophe, but the avoidance of catastrophe, or the silences surrounding it. This obscures but does not *negate* the catastrophe, which his characters deny or refuse to discuss while his readers sculpt its outlines from the scanty details available in his imaginative world. His language contains a Janus-like energy, full of hints and portents that never achieve the transparency of expressed meaning. The evidence of hindsight conspires with the absence of foresight to implicate us in the drama of recognition that constitutes the essence of his art.

Appelfeld crowds his best-known novella, *Badenheim 1939,* with narrative ironies that only an informed reader can pierce. Sometimes they appear as simple declarative statements: "The secret was gradually encompassing the people and there was a vague anxiety in the air, born of a new understanding."[1] These are Tantalus sentences, beckoning one toward an unspecified revelation that never achieves expression. "Alien spirits," "nervous looks," and "distant dreads" infect the text like ripening viruses, spreading symptoms but never naming the disease; we must learn to diagnose as illness what the characters themselves at first mistake for robust health. Echoes of vacancy in the imagery itself urge us to replace absence with content: "The man's voice dripped into him like raindrops pattering into an empty barrel" (25). How much water is required to add substance to empty space? How much insight is needed to give catastrophe a graspable shape?

Omens abound to prompt the guests at Badenheim to ply their sagacity, but they do not relish the clues. They prefer to cling to the self-delusion concealed in bland dialogues like the following, which sound trite but expose with trenchant irony the fatal failure of self-definition that in turn made the victims so vulnerable:

> "I am Austrian born and bred, and the laws of Austria apply to me as long as I live."
> "But you also happen to be a Jew, if I'm not mistaken."
> "A Jew. What does that mean? Perhaps you would be so kind as to tell me what it means?"
> "As far as we're concerned," said Frau Zauberblit, "you can renounce the connection any time you like."
> "That is my argument precisely." (76)

As if to confirm this delusive failure, a moment later a "strange intimacy descended on the dark lobby, an intimacy without words." But for Appelfeld, one overwhelming revelation of the Holocaust experience is the ease with which unspoken intimacies betray, especially when they are based on traditional assurances like the one expressed in the previous dialogue. Intimacies without words lead to unexpected consequences, as we learn a few lines later when the schoolgirl's pregnant condition is disclosed: "The people stood around her looking chastised, as if the facts of life had suddenly given them a slap in the face" (77). The news shatters the twilight hour, but not with the blinding illumination that death as well as life lurks in the future. The so-called facts of life will continue to blur the doom of threatened extinction until the end, inspiring potential victims to substitute visible illusions for invisible threats.

If the unsaid is sayable, is the unseen then seeable? Once more, at least in *Badenheim 1939*, Appelfeld's fiction divides such questions into two verbal tributaries: one irrigating the minds of his readers, the other drowning the vision of his characters. A line like "And the investigations showed what reality was" defines the division between the two and illustrates the twin momentum of Appelfeld's narrative appeal. We the readers are compelled to reenact the split, since we are armed with insight from our own retrospective prescience while simultaneously sharing the failure of foresight *and* insight in the orbit of the characters. When "estrangement, suspicion, and mistrust" begin to invade the town, the characters seem to veer toward *our* orbit; but when we learn that "the people were still preoccupied with their own affairs" (31), we return to the world of the "non-seers." Experiencing the language of the novel through such alternations, we encounter anew the verbal fervor *and* the inert irony of an assurance like "the investigations showed what reality was." This is the challenge facing anyone who enters the realm of the Holocaust: Wherein lies its essential reality, and how do we gain access to it?

Appelfeld's lines have a resonance beyond their immediate definition; as he "says," we are enjoined to "see," but not just what is before our eyes. The impresario Dr. Pappenheim, witnessing the collapse of his tourist season, laments: "If only he had known, he would have designed the whole program differently" (37), and the lament appeals to creators of "reality" at every level: the victim, the reader, the artist, and God himself. Indeed, "If only he had known" becomes the despondent epigraph *and* epitaph for all Appelfeld's fictions about the Holocaust. A bitter regret is one of its major legacies, as the vacationers at Badenheim are to learn, and no small source of that regret, as Appelfeld more than once implies, is the neglect of the resonance native to the word.

This ominous drama is played out repeatedly on the stage we call the text. "You can't imagine the feeling of vitality that a stay in Badenheim gives you," says one of the guests at the resort. "I'm very glad that you decided to stay. It's an event not to be missed." And this brief dialogue ensues:

> "An event?" said Lotte.
> "I can't think of a better word. You're sensitive to words, I see." (41)

A mockery, or a warning? Written with compassion, or scorn? Certainly, exchanges like these are red flags to readers, urging them to consider the possibility—one of Appelfeld's most insistent themes—that alongside the catastrophe to humanity, the Holocaust decreed a catastrophe to language.

In a remarkable story called "Repetition," Austrian writer Peter Handke, alluding to the devaluation of language in the modern era, asks whether it wouldn't be wiser, since words had lost their fairytale magic, "to say that they performed the function of a questionnaire: What is my situation? What is our situation? What is the present situation?" With insouciant cunning, he alludes to the partial source of that devaluation by asking further: "Didn't the term that in the past century designated only 'emigration' lose its innocence now that the events of the last war have changed its meaning to 'forced resettlement'?"[2] Appelfeld's characters in *Badenheim 1939* inhabit a verbal world like this, where the dictionary, as it were, still stands on a shelf in the Garden of Eden. Slogans like "Labor Is Our Life," "The Air in Poland Is Fresher," and "The Development Areas Need You" exert a prelapsarian influence on the unwary guests, who accept customary definitions because they are unaccustomed to a life built on precise analysis of what Handke called "the present situation."

Appelfeld summons us to the avoided task by turning words into vacuums and reminding us bitterly that though nature abhors such environments, human beings at that time apparently did not. He introduces substantives without substance, and makes us wonder why it was so easy to breathe in such unventilated surroundings: "There was a different quality in the air, a sharp clarity which did not come from the local forests" (53). The nouns pile up like empty threats waiting to be charged with meaning: "secret," "anxiety," "understanding," "quality," "clarity." They are like bullets aimed at a target convinced of its immunity, like troops assaulting a Maginot Line persuaded that it is invulnerable.

The odd paradox of Appelfeld's novella about blindness to danger is that the vocabulary prods the reader toward insight, as if the clue to future threats came from the words themselves. If the nouns are barren of meaning, they nonetheless plead for epithets to impregnate them and fulfill their portent. Words begin to link up in a lexical union announcing the "differentness of things," but they occupy an insulated reality: "The light stood still. There was a frozen kind of attentiveness in the air. An alien orange shadow gnawed stealthily at the geranium leaves. The creepers absorbed the bitter, furtive damp" (83). The verbal mosaics seem to resemble familiar patterns, since "frozen," "alien," "stealthily," and "furtive" can be assembled into a kind of descriptive jigsaw puzzle. But in the minds and ears of Appelfeld's characters, they do not interlock; a crucial key is missing. There is no "picture" on the box containing the pieces, reminding us of what they are supposed to look like when they are properly put together. A few lines later, we encounter "huddled," "afraid," and "sudden," but they do not solve the mystery either; they only add to the confusion. More than "discovering" the characters' mistake, the reader *experiences* it: the expectation of pattern, connection, a unified whole is an illusion born of misplaced reliance on the coherence of language itself.

The text is ripe with signs. A musician, told that where they are going he will be a musician still, asks his friend, "In that case, why send us there at all?" This leads to the following exchange:

> The friend sought an impressive formula. "Historical necessity," he said.
>
> "Kill me, I don't understand it. Ordinary common sense can't comprehend it."
>
> "In that case, kill your ordinary common sense and maybe you'll begin to understand." (90)

A central dynamic of *Badenheim 1939* (as well as of other Appelfeld fictions) is the tension between language as "impressive formula" and words as exhausted content. When the famous reciting twins finally perform at the spa, "their mastery was such that the words did not seem like words at all: they were as pure and abstract as if they had never been touched by human mouths." By the end of their performance, "the words did their work alone, flying through the air like birds on fire" (101). Somewhere between language as impressive formula and words untouched by human mouth (and hence ungrasped by human insight) lies that realm of eloquence where the artist forges his sinister implications and verbal ambiguities—the realm, indeed, of Appelfeld's art.

The challenge to the guests at Badenheim—one they fail to meet—is to free themselves from the prison of impressive formulas. Although their lives are in danger, they feel safe between the covers of their trusty thesaurus. "The sun was still shining," says Appelfeld, "but the angry people clung stubbornly to the old words, hoarding them like antiquated gadgets that had gone out of use" (115). Although "words without bodies floated in the lobby," no one recognizes the odd phenomenon—how does one begin? Eventually, "word" becomes one of the novel's characters too, more aware of its displacement, because of Appelfeld's maneuvers, than the characters are of their own. Words know about themselves, and we know about them—that they have lost their roots: "The words did not seem to belong to the present. They were the words of the spring, which somehow lingered on, suspended in the void" (117). Such language is designed to alarm the reader's consciousness; but it lulls the unsuspecting victim in Badenheim into lassitude, until the impulse to sleep and silence replaces the will to speech and significant action.

Both dialogue and commentary contribute to this tension between formula and verbal misunderstanding, as the shadow of Franz Kafka hovers smiling in the wings. "I'm not to blame for anything," cries one of the imminent victims. "It's not a question of crime," replies another (cousin to Joseph K. in *The Trial*), "but of a misunderstanding. We too, to a certain extent, are the victims of a misunderstanding." The world of Kafka infiltrates Appelfeld's novel still further when the narrative voice explains, perhaps too lucidly, "The words *procedure* and *appeal* seemed to satisfy him. He had apparently once studied law. He calmed down a little. The contact with the old words restored him to his sanity" (123). "Crime," "misunderstanding," "appeal"—Kafka has already taught us how futile is the search for these terms in the glossary of atrocity, in a world where victims are unreasonably accused and unseasonably condemned. One of Appelfeld's most remarkable achievements is his ability to adapt the vague threats shaping Kafka's dark vision to the rarer perils of the Holocaust universe. In the absence of a vocabulary to warn, the "old words" soothe their users into a fatally vulnerable state. Is this one of Appelfeld's solutions to the enduring mystery of how it could have happened in the first place?

If the various self-delusions that left the victims unprepared reflect a psychological condition, that condition, Appelfeld insists, was inflamed by the virus of language. Unnerved by the approaching uncertainties, the guests at Badenheim turn on one another, with the former impresario Dr. Pappenheim bearing much of the scorn. As the human alienation intensifies, as the pastry-shop owner declares his own innocence while suggesting that "the east" is the right place for Pappenheim, words too are transformed, their link to meaning crumbling. While the pastry-shop owner vainly invents distinctions between "this" kind of Jew and "that" kind of Jew, "there was a stubborn rhythm in his voice and from the hotel lobby it sounded like someone shouting slogans over a loudspeaker" (142). From the victims' point of view—and this is the one that Appelfeld has always promoted in his writing—here is the culprit. Slogans deaden the feelings of intended victims, even as they "justify" the behavior of the persecutors and nurture indifference in the bystanders. When the "filthy freight cars" appear at the end of *Badenheim 1939,* the victims are left speechless, except for Pappenheim's pathetically desperate remark that dirty coaches must mean that they haven't far to go. The triumph of the enemy signifies the end of separate human identity, as the former guests are "all sucked in as easily as grains of wheat poured into a funnel" (175).

Much of Appelfeld's fiction may be seen as variations on a single theme, as victims and survivors struggle to adapt to the one event forming the next or prior nucleus of their lives: the Holocaust. If *Badenheim 1939* chronicles the doom of an entire community, *Tzili* narrates the fate of one person's estrangement from that community, of a search for contact that never succeeds, of a solitary mind contending with the dilemma of the displaced self. Unlike the guests at Badenheim, Tzili wanders across the countryside *during* the years of the war in a realm where threats abound while forms of defense diminish. As time passes, we endure the effects of the situation sketched in *Badenheim:* the futile and fatal influence of verbal formula on existence.

The myth of the inviolable self is part of the Romantic heritage of Western civilization. Despite his failing health, an ironic reminder of the "dis-ease" implicit in his words, Tzili's father attempts to refute her defects (thereby preserving the myth) with the formula "If you want to you can."[3] It wasn't a judgment, we are told, but a faith, uniting the entire family. Dependence on such formulas slowly thwarted one's ability to function in the world of atrocity. The theme is more insistent in *Tzili* than in any other of Appelfeld's works. Summoned to instruct her in her prayers, a religious tutor asks her "in the traditional, unvarying formula: 'What is man?' " and Tzili replies, "Dust and ashes" (5). But language that once defined human nature and its relation to spirit prompts routine responses here, instead of insight. The catechism about prayer and obeying the commandments of the Torah awakens "loud echoes in Tzili's soul, and their reverberations spread throughout her body" (6); but the description mocks the mystic moment, since Tzili is shortly to be abandoned by her family, left helpless and exposed to a hostile environment.

Slogans that in normal times sustain the unexamined life thread the narrative like hopeful detours to security. "Women are lucky," Tzili remembers having been told. "They don't have to go to war" (56). "Without cigarettes there's no point in living" (73), exclaims Mark, the father of Tzili's unborn child. They are verbal pathways leading to a dead end, language at cross-purposes with the particular reality of the Holocaust, as is stingingly confirmed for us (if not for Tzili) by the "encouraging" words she receives from the nurse when her baby is in fact born dead:

> You must be strong and hold your head high. Don't give yourself away and don't show any feelings. What happened to you could have happened to anyone. You have to forget. It's not a tragedy. You're young and pretty. Don't think about the past. Think about the future. And don't get married. (177–178)

"You have to forget. It's not a tragedy. You're young and pretty." The nurse vanishes from the novel, but her rote reassurances give birth to a voice of their own, having been successfully delivered while the life in Tzili's womb dies. It dies because the language necessary to animate it has been stillborn first. Behind such well-intentioned but misguided formulas cower the despair and futility of the Holocaust-haunted spirit.

If *Tzili* had been designed as a traditional bildungsroman, then Appelfeld's beleaguered adolescent would have discovered a new and vital language to express her private vision of reality. But Appelfeld parodies the form: no sooner does a formulaic principle register itself in Tzili's consciousness (and the reader's too) than its opposite appears, undermining the stability of language, of consciousness, of the very structure of the novel. Beyond mere platitude is the terrifying sense that words no longer match the reality of things, curbing the simplest efforts to understand, to say nothing of defending one's life. When reality steals on you unawares, as it does on Tzili and Mark, what support can one fall back on? "Death isn't as terrible as it seems," says Mark. "A man, after all, is not an insect. All you have to do is overcome your fear" (101). How far have we traveled from Tzili's father's advice that "if you want to you can"?

At first, Tzili is not encouraged by Mark's formula for survival, though he repeats it as he is about to leave to gather provisions: "Once you conquer your fear everything looks different" (103). When he doesn't return on time, his words prop up her failing spirits, giving her back a kind of confidence: "Mark's voice came to her and she heard: 'A man is not an insect. Death isn't as terrible as it seems' " (106). Tzili clings to the promise of Mark's return, though by now we have learned that words offered in conviction need not be true. His refrain becomes a dirge, an ironic epitaph, as it returns in her dreams: "Death is not as terrible as it seems. All you have to do is conquer your fear" (114). The formula achieves a kind of closure when Tzili utters it herself, as a principle of her being, to a group of refugees who apparently have survived an ordeal far worse than hers: "I'm not afraid," she declares. "Death is not as terrible as it seems" (120). One is reminded of Paul Celan's sinister refrain in "Todesfuge": "Der Tod ist ein Meister aus Deutschland." But Tzili has never read Celan or grasped, as he had, how atrocity had rotted the bond joining language to truth.

For a single moment in *Tzili,* it seems as if fact has finally displaced formula, when one of the refugees announces: "Death will follow us all our lives, wherever we go. There'll be no more peace for us" (160). Surely here language has shaped a durable insight, sturdy enough to pass the test of platitude. But the principle of alternating viewpoints that dominates the novel forbids any such dramatic climax as this. Scarcely a page later, the summer sun works its magic and real pleasures return, "as if the years in the camps had vanished without a trace" (163). Onto the screen of reality Appelfeld projects two images, sometimes overlapping, sometimes distinct, but always present: normal existence, with its hopes and illusions ("Death isn't as terrible as it seems. . . . Once you conquer your fear everything looks different"), and the soiling vision of Holocaust fact ("Death will follow us all our lives. . . . There'll be no more peace for us").

Each image imposes part of its contour on the other, so that neither is ever entirely pure. If utopian hopes precede and follow nightmare, nightmare nonetheless casts its shadows in both directions. One day, Tzili is convinced that Mark and the others will return; the next, she realizes that her search and her expectation are in vain. Naïve consciousness rivals a dark history in a never-ending effort to recover the impossible. Oblivion and memory establish a constant flow in the narrative: Shall we reclaim a ruined past, engage in drunken and forgetful revelry in the present, or set out in sober quest of a reborn future? Lacking the inner resources to shape her own life, but still a victim and survivor, Tzili clings to the shreds of promise served up to her by others. "In Palestine everything will be different" (182) is only the last (though far from the strongest) of the formulas—this one containing its share of ambiguity—that thwart her quest for equilibrium.

Some refugees dance; others play cards or sleep; still others choose suicide. Explanations for their behavior only confuse Tzili, who never is able to reconcile language with her experience. A failed bildungsroman (as any novel of the Holocaust must be), an adventure in diseducation, *Tzili* quietly rejects the form in fiction that weds speech to insight and makes character growth the test of a successful imaginative version of reality. Tzili listens to speakers heralding "the agonies of rebirth in Palestine" (182), but the whole momentum of the novel has warned us against such bland formulations. "Birth" in any of its combinations no longer exists outside the orbit of "death"; the Holocaust has cancelled an essential feature of vocabulary, the principle of antonyms.

Therefore, when at the end one of the refugees with Tzili calls out, "We've had enough words. No more words," she acts not as a mouthpiece for Appelfeld, but for a point of view that has been rumbling beneath the crust of the text from the beginning. The Holocaust has taught us to mistrust not only what we hear, but what we have heard. *Tzili* disgorges this meaning despite its lonely and unresponsive "heroine," who like us listens to its ultimate expression from other lips, not, as one might expect, from her own: "I'm declaring a cease-words. It's time for silence now" (182). What appears to have been a narrative of survival turns out to be equally a chronicle of loss: neither exists apart from the other. The refugee who utters these sentiments adds, "Phooey. This rebirth makes me sick" (182), and then takes Tzili under her wing, as if to prepare her for the journey to the Holy Land that may never be a homeland. At least, we are cautioned to be wary of our epithets, lest we betray the ordeal of Tzili and her fellow survivors. This character, about whom we learn only in the novel's closing pages, represents a fusion of attitudes that prepares us to confront the trial she and the others have endured: "She had no regrets. There was a kind of cruel honesty in her brown eyes" (184). The oxymoron will prove surprising only to those who have not relished Appelfeld's scrupulous art of forcing words into complicity with one another.

Cease-words (another oxymoron?) and silence are only a prelude to the cleansing of the cluttered imagination, cluttered by verbal formulas designed for reassurance more than truth, an honesty amputated from cruelty—a worthy aim, perhaps, but one blocked by the shadows that continue to stalk Tzili despite her wish to escape them.

Badenheim 1939 is a story of relocation leading to destruction. *Tzili* is a story of dislocation leading to another kind of relocation. In a sense, at least for our purpose of detecting some progression in Appelfeld's use of the Holocaust in his fiction, *The Immortal Bartfuss* permits a rounding out of the interpretive cycle I have been pursuing. A survivor like Tzili, Bartfuss has been living in the land for which Tzili was about to depart; his fate allows us to test the ambiguous promise that "in Palestine everything will be different."

Is there meaning simply in safe refuge? Has Bartfuss, among Jews, achieved the equilibrium denied to the guests at Badenheim and to Tzili among the refugees? Bartfuss, at least, has experienced the somber tones of Celan's "Death Fugue"; we have his creator's word to confirm it. "The survivor, Bartfuss," said Appelfeld in an interview with Philip Roth,

> has swallowed the Holocaust whole, and he walks about with it in all his limbs. He drinks the "black milk" of the poet Paul Celan, morning, noon and night. He has no advantage over anyone else, but he still hasn't lost his human face. That isn't a great deal, but it's something.[4]

This is a minimal tribute, perhaps, but one must turn to the text to determine how having "swallowed the Holocaust whole" affects one's spiritual digestion.

The reader is obliged to proceed with caution. Bartfuss himself is wary of those who would make him into a legend. Although he internalizes his ordeal, he rarely talks about it. He may have changed, but he will not be changed by others. "They need legends too," he thinks, "heroes, splendid deeds. So they could say, 'There were people like that too. In fact they didn't know a thing about Bartfuss. Bartfuss scrupulously avoided talking about the dark days. Not even a hint."[5] He even resents the label "immortal," which is someone else's invention. He knows how impurities in language can lead to impurities in life. And he is reduced to silence no less by the awe of others than by the scars of his inner wounds. "He withdrew, and words he had once used withered inside him" (61). But the gnawing remnants of his unexpressed past, though muted, have not been silenced; like embers that will not be extinguished, they flare up at odd moments unprovoked. Appelfeld's narrative documents the pain of not being able to share this recurrent inner turmoil.

In a stark reversal of Romantic doctrine, the self has grown inviolable through violation, not through organic union with the world of nature and spirit. Bartfuss will not tolerate further violations. Hence his conviction (mistaken, but understandable, Appelfeld would imply) that "people were born for solitude. Solitude was their only humanity" (100). The loneliness thrust on Tzili by her environment becomes the inner landscape of Bartfuss's life, internalized by him in a way that trades conversations with himself for the social discourse he never can abide. What appear to us as monologues are dialogues to him; words are his companions, and language is the obstacle that prevents him from breaking out of his isolation. This paradox is the secret of his being.

Others may dream of human bonds; Bartfuss does not. Besieged by his loneliness upon waking, he dimly recalls a friendly skirmish with language: "A few words, which he had apparently used in his sleep, skimmed over his tongue. Their warmth still lingered" (68). But the words left over from sleep do not help him in his daily contacts: "years of silence, revulsion, and abstention had brought him close to no one" (70). The two words "in Italy," which ought to unite old acquaintances who once shared the intimacies of this remembered locale, just alienate them further. Bartfuss's few efforts to reawaken memories of the post-Holocaust days "in Italy" meet with only hostility, and when others attempt the same with him, he rebuffs them. One of these encounters ends with Bartfuss angry and his interlocutor withdrawing "like someone caught in the wrong jurisdiction" (74).

In fact, Bartfuss and the other survivors are caught in *two* jurisdictions, and that is their problem. Like Tzili, Bartfuss has his hopeful moments, which rouse him to kinder aspirations than *her* limited imagination is capable of: "Now he would devote himself to the general welfare, he would mingle, inspire faith in people overcome by many disasters. He would no longer think of himself, his agony, but would work for the general good" (75). But at this point he lacks the word "generosity" (which he will wrestle with later) to enable these dim impulses to merge into action. The word "labor," he discovers, smells like distilled alcohol. It is one of those old words on which the Holocaust has cast a shadow. Accordingly, his efforts to translate his new interest in general welfare into genuine intimacy with Schmugler, another survivor, are greeted with chill indifference, and in his frustration Bartfuss watches his words turn into blows.

The other jurisdiction that Bartfuss inhabits promotes different rules and leads to the disjunction isolating him in his prison of the alternately remembered and avoided past. The very language of *Tzili* infiltrates his imagination, the legacy of the Holocaust thus pursuing him through literature as well as through life. For a moment, Bartfuss too turns critic, offering us a retrospective analysis of Tzili's dilemma that is not so far removed from his own: "The people who were in the camps won't betray their obligations. There are sacred debts. A man is not an insect [Mark's very words, in *Tzili*]. The fear of death is no disaster. Only when one has freed himself of that fear can one go forth to freedom" (94). In the process of repeating the language from another tale, Bartfuss reveals himself as a man plagued by a vocabulary he simultaneously longs to, and no longer *can*, believe in. Appelfeld is specific about this inner conflict: "His own words and those of others teemed within him and ravaged him, and for a long time the words rolled about within his brain as if on wheels" (94). A battlefield would be an equally compelling image. How does one pay homage to all those dead without betraying them—or oneself? Like Tzili again, Bartfuss suffers from fatigue because of his embattled verbal situation—he is often on the verge of sleep.

Bartfuss cannot share his knowledge that "man as insect" and "fear of death" are phrases that the Holocaust has defiled beyond usage. Although the Holocaust context of their special demise continues to haunt his memory, in his other "jurisdiction," as a resident of Jaffa in the contemporary world, he needs remnants of that vocabulary to go on living. His mistress Sylvia, another survivor, offers him what she calls a "word for yourself" (101)—"resilience"—from which he recoils, denying in conversation with another the very need he admits through his inner dialogue. But Sylvia's words seem to thaw some of his numbness, since after her death, at her funeral, in a cemetery where "not a word was heard," he suddenly pluralizes his dilemma by asking another mourner, "What have we Holocaust survivors done? Has our great experience changed us at all?" (107). A new word falls unfamiliarly from Bartfuss's lips—"I expect generosity of them," he says (107)—but Appelfeld refuses to permit this apparent moment of conversion to sentimentalize his fiction. Seeds the reader expects to germinate are killed by the fungus that feeds on hope; the next chapter begins: "After Sylvia's death no change took place in his life" (108).

But a kind of resilience has nonetheless taken root in Bartfuss, and the closing chapters of Appelfeld's narrative admit some narrow rays of light into his consuming inner darkness. The survivors he now meets, though they don't all remember *him*, no longer deny that they remember "Italy," a pivotal locale for Bartfuss between the camps and

Palestine, the place where the certainty of death and the possibility of rebirth first contended for primacy in his mind. Does Bartfuss see an image of himself in Clara, who insists that "there really were two Claras, a selfish Clara who wasted most of her money on cosmetics and fashionable dresses; and, alongside her, another Clara, a hidden one, whose heart went out to anyone who came near" (114)? One of the Holocaust's enduring legacies is precisely this principle of the partitioned self, and Bartfuss's recognition of how Clara has managed it nurtures his own resolve to try "to be generous and not stingy" (115).

Is this merely another formula, disguised as a guide to action? On the one hand, Bartfuss himself regards it as "a good slogan" (115); on the other, he seems driven by a need "to be close to the people from whom he had distanced himself" (116). Does this mean his family, from whom he is estranged, or the community of survivors, toward whom he has made tentative gestures throughout the novel? He tries first with his impaired daughter Bridget, but in spite of his goodwill, his efforts to rehabilitate generosity encounter the familiar obstacles of silence and the past. "Words had gone dumb within him," Appelfeld writes, while Bridget's appearance, her dull eyes and full breasts and bewildered expression, only remind him of the passive casual women he had had on the beach in Italy. The reality of "in Italy" intrudes even here, defeating his resolution to improve his behavior as a father.

Nevertheless, Bartfuss *feels* a "strange closeness" to his daughter, though wounded words like "mercy" and "generosity," which emerge from the final pages of the narrative, as well as some that are implied, like "gratitude" and "forgiveness," continue to elude his efforts to wed consciousness to conduct and heal the split within. If in the beginning was the expulsion, the disavowal of community that we saw dramatized in *Badenheim 1939*, the alpha of destruction that soiled the lexicon of human behavior, the omega that will remedy time past, mend the present, and promote a healing future still beckons in the distance. Like the words of love that Hans Castorp whispers as he slogs through the mud on the battlefields of World War I, the Omega watch that Bartfuss has bought for his daughter but not yet presented to her may augur some special meaning for his still unravelling fate. But the final image of the novel is sleep, so though the ironies of art may have gained Appelfeld a piece of tranquillity, we have no such certain comfort for Bartfuss—or for ourselves.[6]

1988

11

Myth and Truth in Cynthia Ozick's "The Shawl" and "Rosa"

he tension between the Holocaust as a chaotic event and our need to find a form that makes it manageable to our imagination continues to hound us, as language veers between truth and myth in its efforts to extract from history a meaning that may not in fact be there.

The myth in Cynthia Ozick's *The Shawl: A Story and Novella* (previously published separately as "The Shawl" and "Rosa") is that what has been lost can be found, like Rosa's lost underpants found wrapped in a towel. The truth is that what has been lost can never be found, partly because (in memory) it has never been lost and thus need not be found, and partly because the loss is so unspeakable and void of ritual that the very terms "lost" and "found," like "wound" and "healing," "trauma" and "recovery," "ordeal" and "survival" are irrelevant to the experience they seek to describe.

Rosa Lublin, whose surname reeks of the deathcamp Majdanek (located in the vicinity of the city of Lublin in Poland), needs no reminders. Her past pursues her in the imagery of her narrative to her present locale of Miami, where "the streets were a furnace, the sun an executioner,"[1] where the beach at night is "littered with bodies," prone, like the victims of Pompeii, but not only Pompeii, "in the volcanic ash," a private beach "locked behind barbed wire," adjacent to a hotel through whose kitchen Rosa flees, past "cooks in their meat-blooded aprons" (47, 48, 49). There *is* an idiom appropriate to Rosa's condition in the novella bearing her name, but it is available only to readers of "The Shawl," where the source of her present dilemma, the meaning of her identity as a "survivor," is sealed. Others pretend to know what Rosa has lived through; she knows what she has died through, what has died in her, what it means to be dead *while* alive. The murder of her daughter, Magda, is a narrative moment, the truth of Rosa's life; the text of memory is everything that follows, allowing the myth to unfold that time heals, that anguish like hers can be forgotten, that a possible future might redeem an impossible past. The myth encroaches on Rosa even as she tries to reject it: to her niece, Stella, she writes in English, an unadapted refugee posing as a successful immigrant; but to her dead daughter, Magda, she "writes" in her native tongue, a "most excellent literary Polish" (14).

The myth is that with patience and sympathy, we can understand Rosa, learn what lay behind her sudden act of vengeance against the secondhand discards of other people's lives when she willfully smashed the "antiques" in her Brooklyn store. Others seem reluctant to explore the latent fury, frustration, or despair that may have driven her to this deed. They are content to make her into a newspaper item, a headline of media slang: "WOMAN AXES OWN BIZ" (18). The truth is that *even with* patience and sympathy, we may never be able to share the legacy of loss that has transformed Rosa into a misplaced disheveled eccentric adrift in the heat of America's Jewish retirement community. "My Warsaw isn't your Warsaw" (19), she reminds the persistent Simon Persky, and Ozick stalks the myth and the truth of this conviction, and its implications, to the very last lines of her narrative.

Is Rosa's return to humanity forever forbidden by the hermetic sources of her woe? Or is this only a dramatic myth parading as cheerless truth? Rosa's life, like her room, is cramped; Persky would liberate her from her prison with his principle of adjustment, which invites us to believe that myths are useful to help us manage unbearable truths, especially when, like him, we are aging and the future looms with limitations instead of possibilities: "For everything there's a bad way of describing," he says, "also a good way. You pick the good way, you get along better" (56). This may be a standard view, but Holocaust memory contends with the *in*describable, and Persky knows little of this.

Rosa knows much. To Persky's principle of adjustment, Rosa replies, "I don't like to give myself lies" (56). An admirable and honest response, but what then of the letters that Rosa writes, or imagines, to her dead daughter? Indeed, what of the language with which Rosa in "The Shawl" envisions Magda's doom, a tiny body hurled against the electrified barbed wire: "She looked like a butterfly touching a silver vine" (9)? Is not the poetry of similitude itself a form of denial, an instinctive stab at inventing Persky's principle at the trauma's very moment of origin? "For everything there's a bad way of describing, also a good way."

But Ozick skillfully refuses to lapse into a mode of mere rivalry. As a former camp inmate, Rosa does not seek to beautify the truth. She reacts candidly to the "slow stinking maroon waterfall that slunk down from the upper bunks [of the barrack], the stink mixed with a bitter fatty floating smoke that greased Rosa's skin" (8–9). This imagery conceals nothing. Similitude is not always poetic; the metaphors portraying Rosa's starving niece Stella are brutally frank: "Her knees were tumors on sticks, her elbows chicken bones" (3). The odor of the barrack, however, fades and grows dim, and Stella lives and puts on weight. But the moment of Magda's murder is different. It does not enter into the chronology of events that have a past, a present, and a future, but possesses a durational integrity that exists outside the flow of normal time.

Some connections need to be restored; others were never severed. Rosa seeks to reenter time and space when she has her telephone reinstalled, but this only reminds her (and us) of the two universes she inhabits. She calls Stella with no apparent news, until her niece exclaims in exasperation, "Rosa, this is long *dis*tance" (64), and suddenly we realize that for Rosa the language of time and space has different meanings, depending on the context she allows herself to hear it in. Long distance at this moment is the open sesame of Rosa's imagination, causing Magda to spring to life with what we would expect to be the truth of her fate, once more come to haunt Rosa. But instead, truth eases into another sort of myth, as the genie of what Rosa later calls the "drudgery of reminiscence" (69) summons up a teenage Magda, ripe with the fruition of what might have been, and Rosa once more enacts, unwittingly, the bitter irony of Persky's well-intentioned maxim: "Life is short, we all got to lie" (56).

If Rosa's great burden is that she has lost Magda in death, an even more searing pain is that she has lost Magda in life. Memory reminds her of the roles she has been deprived of, not only as a potential scientist, but also as mother, as grandmother, as a source of future generations. Her Holocaust history has amputated her from time, robbed her of her future; this is what she means when she tells Persky, "For me there's one time only; there's no after" (58). Persky tries to guide her back into the flow of time, of before, during, and after, by advising, "Sometimes a little forgetting is necessary . . . if you want to get something out of life" (58). In the world of chronology, Persky speaks a usable truth, but in Rosa's universe of duration, dominated by the instant of Magda's murder, the counsel to forget and get on with your life launches a myth both impudent and futile. "Before is a dream," she says. "After is a joke. Only during stays. And to call it a life is a lie" (58).

Is there a way out of this impasse? Persky would perish were he to adopt Rosa's viewpoint and allow his wife's "mental condition" to poison his own life. His wife's disorder, as he describes it—"She's mixed up that she's somebody else. Television stars. Movie actresses" (27)—contains a hint of humor, echoing as it does a common if less constant malady in our media-crazed culture. We can sympathize with the problem of identity confusion in a society that grafts public myth onto personal truth and leaves us uncertain who we really are. But "sympathy" is not a word we can always apply to Rosa's identity confusion *during* her ordeal—that is, what kind of Jew is she?—as she tells of her patrician family's being thrust into the crammed Warsaw ghetto along with "teeming Mockowiczes and Rabinowiczes and Perskys and Finkelsteins, with all their bad-smelling grandfathers and their hordes of feeble children!" (66). Such contempt for her fellow Jews and fellow victims reveals a blind spot that Rosa has never shed, and some of that aristocratic disdain spills over in her initial response to Persky's overtures of friendship. His mediation tests Rosa's ability to shift her identity from one time frame to the other without abandoning the true source of her pain, the indelible memory of Magda's doom.

Stella's strategy, on the contrary, is useless and indeed counterproductive, since it insists on thrusting Rosa into the conventional therapeutic paradigm of illness and recovery, which simply does not fit Rosa's case. By calling for "a new attitude absolutely, recuperated. The end of morbidness" (63), Stella uses a language that deliberately screens Rosa's singular sorrow behind the myth of healing. Some wounds leave a scar whose legacy is an incessant throb, like, in the stunning imagery of the narrative itself, "a terrible beak of light bleeding out a kind of cuneiform on the underside of [Rosa's] brain" (69) as she pens one of her internal letters to her daughter. Letters, like telephone calls, a basic form of communication between human beings through language, have a futility of their own in these stories (the idiotic missives of Dr. James Tree are the most extreme examples), as if Ozick would suggest that words themselves might never pierce the loneliness isolating Rosa or embrace in narrative form the inhuman universe of the Holocaust.

Like Rosa, who is physically immobilized, we are spiritually paralyzed by the Ur-moment of these stories, when "Magda's feathered round head and her pencil legs and balloonish belly and zigzag arms splashed against the fence" (10). Rosa, in terror for her own life, is reduced to silence as she watches her daughter die, swallowing the "wolf's screech" of protest that rises to her lips by stuffing the shawl into her mouth. Narrative can capture the anguished impact of this silence only by admitting its inability to express it. Such silence then enters the realm of memory, where it remains internalized while we as audience hope that time will allow it eventually to be transformed into speech.

But even then, as in Rosa's plight, Holocaust speech as it is uttered can be drowned by other silences. Whether the shawl that once held Magda remains part of Rosa's myth, or her truth, or both, we are left to interpret. When the ardor of art trespasses on the misery of mere experience, it prods consciousness to enter the richer domain of the imagination. Although Holocaust fiction is never totally detached from Holocaust fact—without the SS and the deathcamp, there would have been no stories called "The Shawl" and "Rosa," and indeed, as the narrative hints, there might have been no Magda either—Holocaust fiction and Holocaust fact are not the same. To illustrate this, we might reflect on the differences between them and the considerable value of each by turning to a fragment from a true testimony that might have been—though I know it was not—the seed of Ozick's fiction:

> I had a baby boy. . . . They took us to the buses; they brought us to a big airfield. And nearby were the trains, the cattle trains. And . . . I look back: I think for a while I was in a daze; I didn't know what was happening actually. I saw they [were] taking away the men separate, the children separate, and the women separate. So I had the baby, and I took the coat what I had, the bundle, and I wrapped [it] around the baby and I put it on my left side, because I saw the Germans were saying left or right, and I went through with the baby. But the baby was short of breath, started to choke, and it started to cry, so the German called me back, he said in German, "What do you have there?" Now: I didn't know what to do, because everything was so fast and everything happened so suddenly. I wasn't prepared for it.
>
> To look back, the experience was—I think I was numb, or something happened to me, I don't know. But I wasn't there. And he stretched out his arms I should hand him over the bundle; and I hand him over the bundle. And this is the last time I had the bundle. . . .
>
> But as I look back, I don't think that I had anybody with me. I was alone, within myself. And since that time I think all my life I been alone. [Meanwhile, the camera pans to the other end of the couch where she is sitting, to settle on her present husband—her first one was killed during the Holocaust—whose face is frozen with utter grief and despair.] Even when I met Jack, I didn't tell Jack my past. Jack just find out recently. For me, I was dead. I died, and I didn't want to hear nothing, and I didn't want to know nothing, and I didn't want to talk about it, and I didn't want to admit to myself that this happened to me.

She is deported to Stutthof concentration camp, where she meets the doctor who in the ghetto before her deportation had operated on an infected breast. She continues:

> And when she [the doctor] saw me there she was so happy to see me, and right away she says, "What happened, where's the baby, what happened to the baby?" And right there I said, "What baby?" I said to the doctor, "What baby? I didn't have a baby. I don't know of any baby."

Then she pauses for an instant, nods her head, taps her brow with a finger, and concludes: "That's what it did to me."[2]

Who knows where truth ends and myth begins, or where invention starts to displace—and some might argue misplace—history? Both Rosa and her real-life counterpart have outlived their "logical" destinies—that is, to die like their children in the camps, or to live with them as mothers in a normal world. Deprived of either option, they are trapped in a cocoon of remembering. Shawl or bundle, a shroud is a shroud; they conceal the same doomed child, clutched by the same inconsolable mother. Or do they? Rosa's "My Warsaw isn't your Warsaw" is more than a haughty put-down to distance her from the impertinent Persky; it echoes a difference that divides myth from truth, fact from fiction, chronology from duration, and, most of all, imagination from reality. Real testimony may offer us the concrete grief and tension of the Holocaust experience as no story can possibly do, but Ozick's narratives add a quivering intensity to that experience through vibrations that are unique to the world of fiction. The "Shawled telephone," the final image in "Rosa," presents us with a contradiction in terms, an impossible union between the myth of communication and the isolating truth of Rosa's pain. Her last gesture in the story is to remove the shawl from the phone and to accept its message, the call from Persky, into her life, in an attempt not to replace "then" by "now," but to separate the two and acknowledge that the twin universes she inhabits are distinct. Will this gesture allow her to go on with her life while at the same time preserving her death, and Magda's too? The closing words of "The Shawl" were "Rosa drank Magda's shawl until it dried" (10), while "Rosa" ends with the words "Magda was away" (70), a teasing invitation to the faint of heart to sigh with relief at Rosa's ultimate triumph over her incubus, her poisonous past. The myth beckons. But hardier—and warier—spirits will understand that to be away is not to be gone, that at best Rosa has agreed to an uneasy truce with her legacy of loss. Persky may be here; but Magda will return. The truth endures.

1992

12

Malamud's Jews and the Holocaust Experience

ow much of world literature, from Job and Oedipus through Tolstoy and Dostoevsky to Saul Bellow and Bernard Malamud, embraces the principle that suffering kindles a moral advantage, an inner discipline, a spiritual strength? The capacity to suffer, with its accompanying cleansing and sanctifying of consciousness, distinguishes us from the beasts. It makes us more human. It helps us—theoretically, as lighted by the prism of literature—to bear with dignity the burdens of living and dying. Despite the formulaic tone of these sentiments, they have served the literary imagination as irreducible havens of the human in the midst of the unfolding and expanding oppressions and atrocities of history, especially in our own time. Even a resolute secularist like Camus embraced the disciplinary value of suffering. Undismayed by the spectacle of man adrift in a universe without a master, he could conclude of his Sisyphus that the "struggle itself toward the heights is enough to fill a man's heart,"[1] and of his threatened population in *The Plague* that in time of pestilence "there are more things to admire in men than to despise."[2] Contrary to expectation, and perhaps even experience, we have been taught to believe by the vision of our writers that physical anguish or deprivation inspires the soul and nourishes the quality of our inner lives.

It is not surprising to find that an American writer like Bernard Malamud, concerned with Jewish character and Jewish themes, should fall easily (and admirably) into the pattern of this vision. No matter how bleak the lives of figures like Morris Bober *(The Assistant)* and Yakov Bok *(The Fixer)*, no matter how helpless their natures and hopeless their situations, they retain the gift of suffering, and this keeps their minimal existences from dissolving into despair. Moreover, the gift is transferable and transformational, since Bok's stubborn humanity in the midst of misery infiltrates the sensibilities of one of his jailers, while Bober's finally provides a model of being for his wayward assistant's undisciplined nature. Suffering thus has an exemplary as well as a personal value, and the idea lends literature a resonance that formalists might not applaud but which Malamud has affirmed in novel and story throughout his career.

In an address accepting the National Book Award for *The Magic Barrel*, Malamud was unequivocal about his own humanist position: "I am quite tired of the colossally deceitful devaluation of man in this day. . . ." No one could question the sincerity of this benign attitude, which is not just speech-rhetoric, since a belief in fundamental human dignity pervades Malamud's writing. But it represents a point of view, not a truth, and Malamud's weariness with somber visions that devaluate man cannot validate (though it certainly does much to explain) his truculent use of "deceitful." In a century abounding in war, assassination, mass murder, and acts of terrorism, we are surrounded and almost engulfed by an obvious loss of reverence for the human, at least on the part of the human agents of and participants in these acts. Malamud's conclusion that "the devaluation exists because [man] accepts it without protest"[3] may have some foundation in experience, but it rather naïvely oversimplifies an enormously complicated issue. Charging the enemy with bayonets of verbal dissent does little to illuminate this issue.

Why has a writer like Malamud, in whose work the themes of suffering and dignity are so closely allied, touched so peripherally on the matter of the Holocaust? And how, when he has approached it, has he reconciled its atrocities with his determination to illuminate the human? How does his personal vision accord with the remarkable exclamation of a deathcamp survivor in a videotaped interview: "I saw the sun in Auschwitz, and the sun was black—the sun was destruction"? The extraordinary inversion of imagery, reminding us of Milton's "darkness visible," illuminates not the human but *in*human, and I suspect that even Malamud would find it difficult to declare this implicit devaluation "deceitful." The experience of the deathcamps has transformed many other sources of warmth and light—family bonding, friendship, culture, trust, hope, belief in progress—into perilously vulnerable consolations, an uncongenial possibility that devalues the human whether men protest or not. Such conclusions would naturally appear deceitful to a writer whose characters are snared by their own mistakes, errors of judgment, stubbornness, determination to resist the corruption of their moral nature. But this is no more than confessing that most of Malamud's characters inhabit a universe alien to the premises on which the Holocaust universe are built.

In a fine essay on Malamud's work, devoted chiefly to *The Fixer*, Robert Alter suggests that for Malamud in this novel "1911 is 1943 in small compass and sharp focus," that the Beiliss Case gives him "a way of approaching the European Holocaust on a scale that is imaginable, susceptible of fictional representation." But by reducing the scale of that event from the extermination of a people to the persecution of a person, Malamud does more than make that momentous atrocity manageable; he transforms it into a story of the affirmation of private dignity that elevates the ordeal to tragic dimensions. Alter's very language echoes a vision and a tradition that fall comfortably into a familiar literary stance, but settle uneasily on the circumstantial dilemmas confronting the Holocaust victim. Of Malamud's protagonist in *The Fixer* he says: forced "to summon up all his inner resources of survival in order to stay sane and alive in solitary confinement, Bok in his cell recapitulates the darkest, most heroic aspects of Jewish existence in the diaspora."[4] But how could we use such language to describe that final expression of diaspora in Europe called deportation to the deathcamps?

Alter lucidly describes Malamud's protagonists as "futilely aware of their own limitations . . . 'self-confessed failures' caught in the trap of themselves and rankling over their predicament, though just a little amused by it too."[5] Now imagine writing about protagonists in Holocaust fiction, after the ordeal of the camps, engaging in a self-confrontation resembling this one. Such confrontations are luxuries reserved for sensibilities in situations, painful as they may be, which permit meaningful moral resistance. Bok, for example, will not sign a confession. But who could speak of Treblinka as a "predicament"? Who could be "rankled" by finding oneself in proximity to the gas chamber? What have one's human limitations to do with such a doom? And what constitutes success in such an environment—survival as a starving corpse? And how could one be even a little amused by such a scenario? The critical vocabulary relevant to Malamud's vision can do little to illuminate the problems of character and fate, choice and chance, moral vision and moral failure in a Holocaust setting, since atrocity provides for the writer a perplexing barrage of abnormalities undreamed of in the worst nightmares of a Morris Bober or a Jakov Bok. Death in a place like Auschwitz was neither "human" nor "fate," and though Bober's demise is pitiful, he is in his generosity and folly very much the agent of his own end. Bok's ordeal is a celebration of character over circumstance; fiction about survival in a deathcamp is not a celebration but a concession to a more modest goal: nurturing irreducible needs, minimal gestures to keep the body physically alive.

Despite the ignorant anti-Semitism of the boatman who ferries Yakov Bok from the shtetl to the city, with his augury of the total destruction of the Jews, Malamud's fiction does not anticipate or evoke the horrors of the Holocaust. His ancestry in this novel, beside the pogrom mentality of nineteenth-century Russia, is the prison literature of a Dostoevsky, a Koestler, a Solzhenitsyn, the story of an individual whose moral sensibilities are assaulted by an external oppression which breeds internal resistance in the potential victim. Bok's resolve to flee the shtetl-prison to try his life elsewhere resounds with a deafening irony, but it is a self-chosen flight, and even the arrest and bizarre "justice" he careens toward is an imaginable if painful future, consistent with the atmosphere of pogrom and tsarist oppression. His discourses with himself in prison, his imaginary dialogues at the end with the tsar, also remain in the realm of imaginative possibility. But how would such discourse, private or public in its thrust, shed light on the dilemmas of the concentration camp victim? How would Bok's descendants in the camps carry on with Hitler the kind of conversation that Bok indulges in with the tsar? Would they say to Hitler, as Bok does to Nicholas, that he is lacking "the sort of insight . . . that creates in man charity, respect for the most miserable"? The very vocabulary blushes at its own insolence. What in Holocaust literature could replace the tsar's self-pitying defense: "I am—I can truthfully say—a kind person and love my people. Though the Jews cause me a great deal of trouble, and we must sometimes suppress them to maintain order, believe me, I wish them well."[6] Malamud makes this confrontation the culmination of his novel, imagining the encounter with little difficulty. But how could we imagine Adolf Hitler addressing one of his victims in comparably appropriate tones? George Steiner (in *The Portage to San Cristobal of A.H.*) seems to be the sole contemporary writer to attempt such a feat, and his achievement represents a triumph of rhetoric, not literary art (so did, one might argue, Hitler's). The dilemma of the Jew in Malamud's fiction, in any event, does not foreshadow future artistic challenges, but reflects earlier traditions. The conversation with the tsar in *The Fixer* is not so far removed from Sholom Aleichem's Tevye the milkman's effort to engage his God in dialogue, and his wittily critical tone spills over into the voice of Yakov Bok.

Oppression, suffering, misery, humiliation are *tests* of Jewish character and Jewish dignity in novels like *The Fixer* and *The Assistant*. Life may be a prison for Malamud's Jews, in real jail or in a grocery store from which Morris Bober rarely ventures, but within those limitations much space remains to choose one's future—not only one's attitude, but one's deeds. When Yakov Bok fears that his jailers are poisoning his food, he refuses to eat; his refusal leads to permission to inspect the food in the kitchen and receive his portion there. Equivalent options are unavailable in the imagined world of Holocaust prisons, where starvation was part of the persecutor's design. The premises controlling Jewish existence in each arena are entirely different; and so would be one's treatment of character. By making a specific choice crucial to the remembered action of *Sophie's Choice,* William Styron melodramatizes without illuminating the moral discontinuity representative of the Holocaust experience; by attributing the choice to a Christian prisoner, he skews that experience even further. Christians, whether the Black Hundreds or Ward Minogue and Frankie Alpine, play vital roles in the destiny of Malamud's Jews. But their anti-Semitism afflicts only private fates, providing a mere flickering prelude to the incomprehensible fiery doom that consumed European Jewry decades later.

Circumstances in Malamud always permit moral concern. His Jews are sooner heedless of physical consequences than of this stabilizing and dignifying feature of their inner being. Bok steadfastly refuses to sign a false confession, even if it might lead to his release, because this would implicate his innocent fellow Jews in his alleged crime and impose on him a complicity that would pollute his nature. In his imagined dialogue with the tsar, Bok has his royal disputant sentimentally conclude that suffering has at least taught the prisoner mercy, but Bok promptly rebuts this cliché: "Excuse me, Your Majesty, but what suffering has taught me is the uselessness of suffering." Bok here refers to *unjust* suffering, since his self-imposed suffering, the suffering resulting from his casual indifference to his identity or responsibility as a Jew, has helped to involve him in his present dilemma. His agony has supplemented and clarified Spinoza: "One thing I've learned, he thought," on the novel's final page, "there's no such thing as an unpolitical man, especially a Jew."[7]

Malamud is committed to verifying human identity even in the midst of chaos, which for Yakov Bok signifies amplifying and ultimately affirming the meaning of being a Jew. Living illegally outside the Pale with an assumed Christian name has made him suspect to more than the police. Wrong choices have made him vulnerable. But—to pursue our analogy—wrong choices by the Jew during the Holocaust were meaningless if one was detected, while concealing one's identity successfully was one of the surest means of survival. These circumstances had nothing to do with moral concern, since even the minimal justice available to a Yakov Bok was inaccessible to the Holocaust victim. If the old terminology lingers in discussions of fictional treatments of that event, it may be because as readers we have been trained to view literature through its predecessors and their moral assumptions. For these assert continuity in moral vision despite the disruptions of the Holocaust. Quick to trust, slow to condemn, unwilling to hate, Morris Bober in *The Assistant* pays with his life even during his life for his belief in the value of suffering. And he becomes thereby a shining example of goodness to his assistant, former hoodlum, former anti-Semite, finally a convert to the Judaism of mutual responsibility that Morris clings to in his world of diminished possibilities.

Once again, the vocabulary is appropriate to Malamud's moral vision; but can *The Assistant,* any more than *The Fixer,* cast a light on the shrunken human possibilities of the Holocaust experience? At least one reader has insisted that Malamud means the book "in part, to be a metaphor for the Holocaust experience." In an essay on Malamud called "Metaphor for Holocaust and Holocaust as Metaphor," Michael Brown follows Robert Alter in arguing that the Holocaust is symbolic of the modern condition. He goes further: the Holocaust is the ultimate means "that modernity has invented to destroy man by making his environment inhuman and by degrading his person." Universalizing the event leads to the conclusion, as Brown says approvingly, that there is a connection "between the struggles of American blacks, Russian Jews, and many others for the right to be themselves, and the sufferings of Holocaust Jews because of what they were." But whose advantage do we serve by likening inferior housing and job prejudice among minorities in America, or government hostility to religious practices and refusal of permission to emigrate in the Soviet Union, to the ordeal of Jews in Auschwitz and Maidanek? Or by suggesting, as Brown does, that the gas radiator which nearly killed Morris Bober, whether through accident or intention, is "surely a reference to the fate of Jews in Auschwitz and other camps"?[8]

The Assistant, written before *The Fixer* but set chronologically after its time period, during the very years, in fact, when Hitler was consolidating his power, is an obvious quarry for anyone determined to mine the Holocaust from its pages. Morris Bober, that island of Jewish stolidity amid reefs of anti-Semitic assaults, is a natural candidate for the prototypical representative of the Holocaust victim. Such a view, however, deflects us from Malamud's own central concern, which is Frankie Alpine's efforts to understand what makes Morris "tick," as man and Jew. The novel, after all, is called *The Assistant,* not *The Grocer,* and the main character conflict, between desire and love, force and friendship, guilt and repentance, pilfering and honesty, consistency and self-contradiction, is embodied in Frankie, not Morris. Morris we pity, Morris we admire; but Frankie we need to understand, as he needs to understand himself. Malamud is unequivocal here: Morris the Jew need learn nothing about the fraternal impulse, which he possesses as a birthright. He suffers for others. It is his fate; it is also, ironically, his undoing.

If *The Fixer* dramatizes the dilemma of the Jew as a result of nineteenth-century pogrom mentalities, *The Assistant* presents the Jew as victim in the twentieth century of different forms of anti-Semitism, unconnected to judicial procedures, however phony. Against Ward Minogue's brutal, ignorant, open violence Morris is defenseless; the few glimpses we have of this familiar but frightening figure of the terrorist without conscience makes us wish that Malamud had not dispensed with him so conveniently. We will never understand the Holocaust until we understand its Ward Minogues. But Frankie Alpine's anti-Semitism is more subtle, though no less familiar, than Ward's; it is, however, complicated by twitches of remorse that occasionally awaken echoes of a conscience smothered or stillborn. Frankie simply cannot penetrate the motives for Morris's way of life, and this breeds contempt: "What kind of man did you have to be born to shut yourself up in an overgrown coffin. . . . The answer wasn't hard to say—you had to be a Jew. They were born prisoners."[9] There is neither need nor justification for finding in Frankie Alpine's attitude here the mentality of the latent SS man, or for seeing in "prisoners" and "coffin" antecedents of deathcamp and crematorium. Here, if anywhere, Malamud universalizes the predicament of the dispossessed like Frankie, searching for a core of goodness while rationalizing his failures through hatred and exploitation of others. The one place where Malamud possibly *does* shed some light on Holocaust themes is in his representation of those who use the Jew to conceal their own limitations as human beings.

Morris may be overwhelmed at the end by the waste that has been his life, but we are touched by the consideration that has been his heart. What happens next is that many readers, guided by an instinctive desire to transform the Holocaust ordeal into a heroic challenge, supported by critical enthusiasm for the same end, accept Morris Bober's behavior as a kind of existence exemplary of the camp experience. When Michael Brown writes of Morris that he "suffers honorably and quietly in order to maintain his Jewishness, his *menschlichkeit,* in a world that seeks to obliterate everything Judaism stands for," he transgresses the bounds of Morris's moral vision, and perhaps even Malamud's. Morris did not inhabit such a world, though European Jewry did. But by speaking of Morris as a victim "who knows what he suffers for and accepts the burden,"[10] the critic furnishes us with an easily assimilable, admirable profile of the Holocaust victim too, who bore his burden with honor and dignity and like Morris Bober may have paid for them with his life, but never left in question that he was worthy of his sufferings and his sufferings worthy of him.

But literary tragedy, however modified (as in *The Assistant*) by a character's restraint, is not simply transferable to historical atrocity. Whatever Morris's burdens, he does not face slow starvation, exhaustion, daily beatings, untreated disease; he does not spend his nights picking lice from his scalp and the crevices of his clothing; he does not live in proximity to gas chamber and crematorium, redolent of extermination. How can we consider this kind of agony exemplary? To convert the focus or the scope of suffering in a tsarist prison or a failing Brooklyn grocery store into metaphors of the Jewish experience of atrocity in the concentration camps is to abuse metaphor and to distort one-half of the comparison. Certainly it is consoling for most of us to make inhuman conditions (and, often, less than "human" responses to them) appear more human; but that is an expression of our needs, not of the conditions we recoil from. Only a search for such consolation could lead someone to describe the Holocaust, as Brown does, as "a potent symbol of the dangers posed by the modern world to any person who would be different, to anyone who would insist on being a man."[11] Most Holocaust victims simply hoped to remain alive, nor did they see anything symbolic in what threatened to consume them.

Malamud himself is not guilty of such abstract assertions, though his conventional moral vision may unintentionally encourage them. There are terrible griefs, he knows, and there are inexpressible ones, and he does not confuse them. When he uses metaphors for the Holocaust, which he does more rarely than some of his critics insist, he keeps his distinctions clear. In the brief tale "The Loan" from *The Magic Barrel,* Lieb, his friend Kobotsky, and his wife Bessie all have their lost youths to mourn, with equal legitimacy. Jews have had a hard life in the twentieth century. But the miseries of Malamud's triumvirate are set into grim context when Bessie suddenly sniffs the burning bread, wrenches open the oven, and is greeted by a "metaphor" that reminds us unforgettably of the relativity of woe: "A cloud of smoke billowed out at her. The loaves in the trays were blackened bricks—charred corpses."[12] Does Malamud use this metaphor as a warning against Bessie's stifled charity, as an invitation to the rebirth of fraternal feeling from the ashes of Auschwitz? Or does he suggest that Bessie's Holocaust background has altered in her the spirit of charity that comes so easily to a Morris Bober? The didactic reading would neutralize the impact of the tale; the gloomier one finds some support from the two longer stories in *The Magic Barrel* that use authentic Holocaust survivors to dramatize their themes.

"The Lady of the Lake" *is* a warning, one so transparent that it diminishes the story's effect. Henry Levin, who has repudiated his Jewish identity and assumed the name Henry R. Freeman, is tired of the limitations imposed on him by the past. This sentiment from the opening paragraph meets its mirror image in the closing lines when the Italian girl whom Levin wants to marry turns out to be a Jewish survivor of Buchenwald, who declares: "I can't marry you. We are Jews. My past is meaningful to me. I treasure what I suffered for."[13] Levin, who has denied to the girl that he is Jewish, is trapped by his own lie, and loses the girl. But we have here only the confrontation of two attitudes, not two human beings, since Levin's boredom with ancient history (including, presumably, the Holocaust) is as brittle and unconvincing as Isabella's survivor status, confirmed by numbers tattooed on the unlikely, not to say impossible, area of her breasts. Her pitiful emblem, certainly not metaphor, of her Holocaust experience cannot adequately project that experience, which remains locked behind the barriers of mere verbal assertion.

Malamud realizes more fully the dilemma of confrontation between Holocaust survivor and the vain, insensitive American Jew in his finest story using the Holocaust theme, "The Last Mohican." But even here, the focus is not on the survivor, Susskind, whose past is important in the fiction only as it impinges on the sensibilities of Fidelman, the American. And once again, unlike other Holocaust writers, Malamud is not concerned with evoking particular atrocities but with the failure or inability to picture a kind of suffering that might penetrate and undermine our pompous façades of self-assurance. "Imagine all that history," exclaims Fidelman as he arrives in Rome and contemplates the Baths of Diocletian. But he has a more recent history to "imagine," the history that has reduced Susskind to an impoverished conman without funds, homeland, passport, or even meaningful identity. And this, for Malamud, is the fundamental challenge of the Holocaust, distinct from the romantic appeal that inspires Fidelman and insulates him from unpleasant truths: "History was mysterious, the remembrance of things unknown, in a way burdensome, in a way a sensuous experience. It uplifted and depressed, why he did not know, except that it excited his thoughts more than he thought good for him."[14] Fidelman lacks a language for understanding Susskind's past, and hence his own. As long as he depends on such conventional formulas, history, the past, his Jewish heritage, and the Holocaust experience will remain mysteries to him that have nothing to do with his obscure definitions. His search for Susskind through the ghetto, in the synagogue, to the Jewish cemetery, is a voyage of discovery, but until the very end he does not comprehend the signs that attempt to guide him. When the beadle in the synagogue mentions to the inquiring Fidelman, "My own son—killed in the Ardeatine Caves," Fidelman murmurs his regrets, but surely has never heard of the place or of the infamous Nazi atrocity that occurred there. And when his search takes him to the Jewish section of the cemetery, where Susskind sometimes works, Fidelman gazes at a monument inscribed "For my beloved father / . . . Murdered at Auschwitz by the barbarous Nazis / " but concludes with disappointment: "But no Susskind."[15] An adequate response, a true insight, would have been "But *yes* Susskind," since in this testament to Holocaust victims lay the history of his antagonist, which would explain his behavior and in turn mirror the insufficiencies of Fidelman's life. But he is still blind to the necessity for such insight.

A dream vision leads to the transformative experience in Fidelman, a vision combining Christian and Jewish heritage with the fundamental question of Fidelman's vocation and Malamud's career: "Why is art?" He dreams of a Giotto fresco of St. Francis giving his garment to a poor knight, and intuitively learns what he must do—surrender to Susskind the suit that the survivor needs more than he does. Whether the waking "triumphant insight" that Fidelman arrives at corresponds exactly to the dream vision of charity, of the basic human gesture, the reader is left to interpret, since Fidelman's last words to Susskind, "All is forgiven," remain ambivalent. But Susskind's final words to Fidelman—"The words were there, but the spirit was missing,"[16]—are less ambiguous, and they lead directly to a final evaluation of Malamud's encounter with the Holocaust theme and some of the meanings that have been imposed on that encounter.

Malamud's involvement with the Holocaust has been minimal, possibly because he realized how uncongenial its atrocities were with his impatience at the modern devaluation of man. Facing the difficulty of translating such inhumanity into artistic vision, the writer exposes himself to the charge that the words are there but the spirit is missing, though not necessarily with the derogatory intent of Susskind's accusation. Holocaust atrocity is often not commensurate with spirit of any kind. Malamud critics like Michael Brown, however, noting accurately Malamud's desire to portray men in circumstances allowing them to achieve their potential humanity to the full, extend Malamud's limited ambitions toward the Holocaust beyond verifiable frontiers, ones certainly unverifiable in Malamud's fiction: "The inhumanity of the modern condition can be defeated, as the Jews defeated the Holocaust through their survival."[17] I suspect that Malamud himself would be astounded to find this principle deduced from his novels and stories about particular Jews affirming dignity despite their suffering. Immodest claims like these aggrandize his achievement without illuminating it. Malamud's indirect, tentative, circumscribed inroads on Holocaust reality leave untouched vast areas of harsh and unbearable experience that require fresh explorations of the conventional bond linking the word and the spirit. The results may be more frightening than Malamud's, but no less responsive and significant than his own to the cogent question which he himself raises: "Why is art?"

1987

13

The Americanization of the Holocaust on Stage and Screen

e bring to the imaginative experience of the Holocaust a foreknowledge of man's doom. Not his fate, but his doom. The Greeks sat spellbound in their arenas in Athens and witnessed the unfolding of what they already knew: proud and defiant men and women submitting to an insurrection in their spirit that rebelled against limitations. Oedipus and Phaedra, Orestes and Antigone hurl their own natures against laws human or divine, suffer the intrusions of chance and coincidence, but *make their fate* by pursuing or being driven by weaknesses or strengths that are expressions of the human will. Whether they survive or die, they affirm the painful, exultant feeling of being human; they declare that man, in the moral world at least, is an agent in the fate we call his death.

But the doom we call extermination is another matter. The Athenians could identify the death of their heroes on the stage with a ritual for renewal, ally tragedy with comedy, and make both a cause for celebration. The human drama allowed it. But the Holocaust presents us with the spectacle of an inhuman drama: we sit in the audience and witness the unfolding of what we will never "know," even though the tales are already history. The tradition of fate encourages identification: we may not achieve the stature of an Oedipus or a Phaedra, but their problems of identity, of passion, of moral courage, of retribution, are human—are ours. The tradition of doom—a fate, one might say, imposed on man by other men against his will, without his agency—forbids identification: For who can share the last gasp of the victim of annihilation, whose innocence so totally dissevers him from his end? We lack the psychological, emotional, and even intellectual powers to participate in a ritual that celebrates *such* a demise. We feel alien, not akin. The drama of fate reminds us that man, should he so choose, can die for something; the drama of doom, the history of the Holocaust, reveals that whether they chose or not, men died for nothing.

This is not a comfortable theme for the artist to develop, or for an audience to absorb. Traditions of heroic enterprise, in literature or in life; conceptions of the human spirit, secular or divine; patterns for imagining reality, whether written or oral—all have prepared us to view individual men and women in a familiar way. Hence it should not be surprising that some of the best known attempts to bring the Holocaust theme to the American stage—Frances Goodrich and Albert Hackett's *The Diary of Anne Frank,* Millard Lampell's *The Wall,* and Arthur Miller's *Incident at Vichy*—as well as films like *Judgment at Nuremberg* and the TV "epic" *Holocaust,* should draw on old forms to reassert man's fate instead of new ones to help us appreciate his doom. To be sure, visually we have progressed in thirty years from the moderate misery of a little room in Amsterdam to execution pits and peepholes into the gas chambers of Auschwitz in *Holocaust;* but imaginatively, most of these works still cling valiantly to the illusion that the Nazi genocide of nearly 6 million human beings has not substantially altered our vision of human dignity. When Conrad's Marlow in *Heart of Darkness* returns from the Congo to speak with Kurtz's Intended, he brings a message about Kurtz's inhuman doom to a woman who wishes only to hear about his human fate. And Marlow submits: the truth "would have been too dark—too dark altogether. . . ."

How much darkness must we acknowledge before we will be able to confess that the Holocaust story cannot be told in terms of heroic dignity, moral courage, and the triumph of the human spirit in adversity? Those words adhere like burrs to the back of a patient beast, who lacks the energy or desire to flick them away lest in doing so he disturb his tranquillity. Kurtz's Intended pleads with Marlow for "something—something—to—live with." The Holocaust—alas!—provides us with only something to die with, something from those who died with nothing left to give. There is no final solace, no redeeming truth, no hope that so many millions may not have died in vain. They have. But the American vision of the Holocaust, in the works under consideration here, continues to insist that they have not, trying to parlay hope, sacrifice, justice, and the future into a victory that will mitigate despair. Perhaps it is characteristically American, perhaps merely human, but these works share a deafness (in varying degrees) to those other words that Conrad's Marlow brings back only to find that he has no audience prepared to listen: " 'Don't you hear them?' The dusk was repeating them in a persistent whisper all around us, in a whisper that seemed to swell menacingly like the first whisper of a rising wind. 'The horror! The horror!' "

There is little horror in the stage version of *The Diary of Anne Frank;*

there is very little in the original *Diary* itself. Perhaps this is one source of their appeal: they permit the imagination to cope with the idea of the Holocaust without forcing a confrontation with its grim details. Like the *Diary*, the play (though even more so) gives us only the bearable part of the story of Anne and the other occupants of the secret annex; the unbearable part begins after the final curtain falls and ends in Auschwitz and Bergen-Belsen. An audience coming to this play in 1955, only a decade after the event, would find little to threaten their psychological or emotional security. No one dies, and the inhabitants of the annex endure minimal suffering. The play really celebrates the struggle for harmony in the midst of impending disruption, thus supporting those values which the viewer instinctively hopes to find affirmed on the stage. To be sure, in the *Diary*, Anne is not oblivious to the doom of the Jews, despite her limited access to information; but there is no hint in the play of this entry from October 9, 1942: "If it is as bad as this in Holland whatever will it be like in the distant and barbarous regions [the Jews] are sent to? We assume that most of them are murdered. The English radio speaks of their being gassed."[1] In the *Diary*, however, Anne does not brood on the prospects of annihilation; she devotes most of her reflections to her aspirations as a writer and her passage through adolescence and puberty to young womanhood. Nevertheless, a certain amount of ambiguity lingers in her young mind (absent from her character in the play) that at least adds some complexity to her youthful vision. "I see the world being turned into a wilderness," she writes, "I hear the ever approaching thunder, which will destroy us too, I can feel the sufferings of millions and yet, if I look up into the heavens, I think that it will all come right, that this cruelty too will end, and that peace and tranquillity will return again."[2] But for all but one of the inhabitants of the annex, nothing came right, cruelty grew worse, and neither peace nor tranquillity ever returned.

Yet this is not the feeling we are left with in the play, which accents Anne's mercurial optimism at the expense of the encroaching doom that finally engulfed them all. Upbeat endings seem to be de rigueur for the American imagination, which traditionally buries its tragedies and lets them fester in the shadow of forgetfulness. The drama begins with Otto Frank, a "bitter old man," returning to the secret annex after the war and finding that Anne's diary has been preserved. His "reading" of excerpts becomes the substance of the play, which after the discovery and arrest fades back into the present, revealing a calm Otto Frank, his bitterness gone. Considering the numerous "last glimpses" of Anne we might have received from this epilogue—one eyewitness in Bergen-Belsen, where she died, described her like this: "She was in rags. I saw her emaciated, sunken face in the darkness. Her eyes were very large"[3]—one wonders at the stubborn, almost perverse insistence in the play on an affirmative epigraph, almost a denial of Anne's doom. Why should the authors think it important that we hear from Otto Frank, in almost the last words of the play, the following tribute, *even if those words were quoted verbatim from Anne's real father:* "It seems strange to say this," muses Frank, "that anyone could be happy in a concentration camp. But Anne was happy in Holland where they first took us [Westerbork detention camp]."[4]

The authors of the dramatic version of Anne Frank's *Diary* lacked the artistic will—or courage—to leave their audiences overwhelmed by the feeling that Anne's bright spirit was extinguished, that Anne, together with millions of others, was killed simply because she was Jewish, and for no other reason. This theme lurks on the play's periphery, but never emerges into the foreground, though one gets a vague hint during the Hanukkah celebration that ends Act I. That Anne herself, had she survived, would have been equal to this challenge is suggested by her brief description of a roundup of Amsterdam Jews witnessed from her attic window:

> In the evenings when it's dark, I often see rows of good, innocent people accompanied by crying children, walking on and on, in charge of a couple of Germans, bullied and knocked about until they almost drop. No one is spared—old people, babies, expectant mothers, the sick—each and all join in the march of death.[5]

But the audience in the theater is sheltered from this somber vision, lest it disrupt the mood of carefully orchestrated faith in human nature that swells into a crescendo just before the play's climax, when the Gestapo and Green Police arrive to arrest the inhabitants of the annex. One is forced to contemplate Anne's restive intelligence at its most simple-minded, as Goodrich and Hackett have her reply to Peter Van Daan's irritable impatience at their dilemma with the pitiful cliché: "We're not the only people that've had to suffer. There've always been people that've had to. . . ."[6] Anne's mind was more capacious, if still undeveloped, but a probe into the darker realms that Conrad and Marlow knew of, an entry like the following from Anne's *Diary,* would have introduced a discordant note into the crescendo I have mentioned:

> There's in people simply an urge to destroy, an urge to kill, to murder and rage, and until all mankind, without exception, undergoes a great change, wars will be waged, everything that has been built up, culti-

> vated, and grown will be destroyed and disfigured, after which mankind will have to begin all over again.[7]

This view of the apocalypse before any fresh resurrection appears nowhere in the stage version of Anne's *Diary*. Indeed, its presence in the other works I will examine will be one test of their authenticity as Holocaust literature. If in the end even Anne Frank retreated to a safer cheerfulness, we need to remember that she was not yet fifteen when she wrote that passage. The line that concludes her play, floating over the audience like a benediction assuring grace after momentary gloom, is the least appropriate epitaph conceivable for the millions of victims and thousands of survivors of Nazi genocide: "in spite of everything, I still believe that people are really good at heart." Those who permit such heartwarming terms to insulate them against the blood-chilling events they belie need to recall that they were written by a teenager who could also say of her situation: "I have often been downcast, but never in despair; I regard our hiding as a dangerous adventure, romantic and interesting at the same time."[8] Her strong sentimental strain, which was only part of her nature, dominates the drama, and ultimately diverts the audience's attention from the sanguinary to the sanguine, causing them to forget that the roots are identical, and that during the Holocaust man's hope was stained by a blood more indelible than the imaginary spot so distressing to Lady Macbeth. By sparing us the imaginative ordeal of such consanguinity, the drama of *The Diary of Anne Frank* cannot begin to evoke the doom that eventually denied the annex's victims the dignity of human choice.

The play presents instead a drama of domestic pathos; it begins and ends with the figure of Otto Frank, a paterfamilias without a family who nevertheless is inspired, like the rest of us, by his dead daughter's steadfast devotion to hope. Bruno Bettelheim's needlessly harsh criticism of the Frank family for failing to recognize the crisis for Jews in Europe and to increase the prospect of survival by seeking separate hiding places nevertheless implies an important truth for anyone seeking to portray the Holocaust experience with insight. The family unit, that traditional bulwark in moments of familiar stress, was worthless and occasionally injurious to individual survival in the unpredictable atmosphere of the deathcamp. The tensions that sundered such ancient loyalties are absent from *Anne Frank;* they begin to appear in Millard Lampell's play *The Wall* (1960), based on John Hersey's novel, but even here, under the pressures of life in the Warsaw ghetto, family unity finally asserts itself and triumphs over the strains that threaten to crack it.

The American imagination seems reluctant to take the non-Kierkegaardian leap into unfaith that might reveal a vision like the following, from the Auschwitz stories, *This Way for the Gas, Ladies and Gentlemen,* of the Polish survivor Tadeusz Borowski:

> Here is a woman—she walks quickly, but tries to appear calm. A small child with a pink cherub's face runs after her and, unable to keep up, stretches out his little arms and cries: "Mama! Mama!"
>
> "Pick up your child, woman!"
>
> "It's not mine, sir, not mine!" she shouts hysterically and runs on, covering her face with her hands. She wants to hide, she wants to reach those who will not ride the trucks, those who will go on foot, those who will stay alive. She is young, healthy, good-looking, she wants to live.
>
> But the child runs after her, wailing loudly: "Mama, mama, don't leave me!"
>
> "It's not mine, not mine, no!"[9]

One has only to immerse oneself in this situation to understand how thoroughly the Nazi system of terror and genocide poisoned that vital source of human dignity that made man an instrument in his fate: the phenomenon of choice. Mother and child is a comforting image when the mother can do something to comfort her child; but how does one comfort her child when both are on their way to the gas chamber? Futility drives the mother in this fictional passage—though we have eyewitness testimony to prove the historical bases of moments like these—to a repudiation unthinkable for the civilized mind. But Auschwitz introduced the realm of the unthinkable into the human drama, and no representation of the Holocaust that ignores this realm can be considered complete.

Lampell's *The Wall* peers into its dark recesses, but finally withdraws to reassert a familiar moral view. It is too dark—too dark altogether. "Man as a helpless victim," writes Lampell in his introduction to the play. "I do not deny that this is a truth of our time. But it is only one truth. There are others. There is understanding, and indomitable faith, and the rare, exultant moments when one human finally reaches out to accept another." But can one truth be severed from the other? And how do those rare exultant moments affect the doom of the 6 million helpless victims who did not survive to appreciate them? Lampell chose to avoid this question by searching for flickering rays amid the brooding gloom. His initial response to the plan to dramatize the Hersey novel was to immerse himself in documents and writings about the Warsaw ghetto. When he discovered—to his dismay—that "what was unique in Warsaw was the scope of man's inhumanity to man," he was reduced to "an overwhelming sense of ashes and agony" and almost abandoned the project. His reason would be ludicrous, were the subject not so grave: "I am a writer chiefly concerned with life, not death." By temporarily retreating in despair from the ruins of the ghetto because "I simply could not recognize human life as I knew it," Lampell unwittingly refused the higher challenge of the Holocaust experience—its utter transformation of human life as we know it. By finally accepting the "lower challenge" and building his play about the safer theme of men "in spite of it all, stumbling toward a possible dignity," he still writes an honest play, better theater than *The Diary of Anne Frank*, but one not governed by the inner momentum of the Holocaust toward extermination. Instead, Lampell restores to men an instrumentality in their fate; a handful of Jews, he insists, "exposed the fullest potential of the human race."[10] Whether this is solace for the 300,000 other Jews of the ghetto who were deported to Treblinka (not Auschwitz, as the play suggests) and murdered, Lampell does not consider; but the impact of the dramatic spectacle is to affirm the heroic fate of the few, and to mute the unmanageable doom of the wretched rest.

The consolation of the Hanukkah celebration at the midpoint of *The Diary of Anne Frank* has its analogue at the exact midpoint of *The Wall:* the wedding of Mordecai and Rutka. But Lampell has established an effective tension between joy and terror, for while the wedding ceremony proceeds indoors, the first roundups of Jews for deportation to "labor camps" occur in the streets outside. The counterpoint between the grave Chassidic dance of Reb Mazur and the confused screams of terrified Jews fleeing for their lives in disorder, between the lively hora of the celebrants and the thudding boots of the Nazis, defines two alienated realms—and suggests the futility of the Jews' trying to inhabit both. Bruno Bettelheim had charged that the Franks' refusal to abandon family ritual and tradition had cost them their lives, though he seemed unable to admit that after twenty-five months of security in their haven, with the Allies moving on Paris and soon to be in Belgium, the Franks had sufficient reason to believe that their strategy for survival would work. Lampell assumes a more complex view, acknowledging how insidiously the Nazi threat of extermination could infiltrate the family unit itself. The desperate dependence on stability represents almost willful blindness on the part of the ghetto inhabitants, and to his credit, Lampell makes this "refusal to see" the dramatic focus of the second half of the play.

What it amounts to, for the historic individual as well as the dramatic character, for the artist, the reader, and the audience, is accepting the credibility of doom (extermination in the gas chamber for being a Jew) when all our lives we have struggled to absorb the painful truth of our mortality, "merely" the necessity for our death—man's fate. Against that fate we can mount consolations, and even some forms of transcendence—faith, love, children, creative endeavor, some communion with the future that liberates us from the prison of our mortality. But what promised to free the human spirit from the deathcamp, the gas chamber, and the crematorium? This is the question to which Lampell is committed—*before* the act of creation—to finding an affirmative answer. Such commitment requires him to manipulate probability and distort the balance between heroism and despair, as if a prior espousal of human community, even amid the rubble of the destroyed ghetto, were the only way to make the Holocaust acceptable on the stage. To be sure, violations of that sense of community appear in the play. The paterfamilias here, Pan Apt, the father of Rachel, Halinka, David, and Mordecai, repudiates his Jewish heritage and with forged papers flees to the Polish side of the Wall, deserting his family. Stefan, Reb Mazur's son, joins the Jewish ghetto police because of the Nazi promise that he and his family will be safe from deportations. But in order to save himself, he is reduced to pleading with his own father to consent to "resettlement." Nazi doom left the individual no simple way of surviving with dignity: one had to pay a human price for his life, and in *The Wall,* that price is usually a disruption of family integrity. The one place where such integrity *is* preserved is one of the least probable (though most reassuring) moments of the play, infringing on the authenticity of history and of human doom in the Holocaust. Huddled in their bunker as Nazi troops are rooting out the scattered remnants of resistance, Mordecai, Rutka, their baby (born in the ghetto), and a few other underground fighters are about to make their way through the sewers to safety outside the Wall, when a detachment of German soldiers spots them. Withdrawing into the bunker again, they wait in silence, when suddenly the baby begins to cry. We have ample evidence of parents smothering their infants to protect themselves and often larger groups from detection, but Lampell flinches before this ultimate rejection of the family bond: it, too, would have been too dark—too dark altogether. His scene direction reads: "Mordecai puts his hand over the baby's mouth. Rutka stands it as long as she can, then shoves his hand away."[11] The baby cries again, and of course the Germans hear him. Verisimilitude, human as well as aesthetic truth, would require the capture of the Jews at this point; instead, in an utterly unconvincing denouement, Berson leaves the bunker undetected and draws the soldiers away with the sound of his concertina. The heroic impulse triumphs over truth, Berson sacrifices himself that others may live (assuming they do: otherwise, his gesture would be doubly futile), and man proves himself still in control of his fate, willing to surrender the life of a man in order to assure the continuity of the Life of Men.

One of the paradoxes of this play, and of much Holocaust literature, is that it tries to serve two masters. When Rachel reports to her horrified fellow Jews about evidence that the deported victims—men, women, and children—are not being resettled, but murdered, they are unable to believe her. "It doesn't make sense,' cries Reb Mazur. "All right, they're not civilized," objects Mordecai, her brother. "But they're still human." And the rabbi later affirms: "Yes, I am calm—because I know that any faith based on love and respect will outlive any faith based on murder." One "master" which this play serves is a familiar but weary vocabulary, words like "sense," "civilized," "human," "faith," "love," and "respect," which desperate men and women cling to, to shore up the ruins of their crumbling world. They lead finally to the heroic if ultimately futile resistance that concludes the play—futile because the Jews possess neither numbers nor weapons sufficient to defeat the Germans. The other "master," mouthed but not acted upon by the briefly cynical Berson, represents a subterranean truth of the Holocaust that art has not yet found an adequate vocabulary to explore: "Love. What does it mean? When they come with the guns, I have seen a son beat his own mother. A mother throw her child out of the window, her own child that she bore inside her." We get a glimpse later on of the motive that leads Berson to a moral reversal, bringing him back into the ghetto after he has left it: "I always thought that just to live was enough. To live *how?* To live *with whom?*"[12] But fine sentiments cannot replace the searching psychological analyses art must provide us with if we are to understand how the human creature reacts when he realizes that his enemy is neither "civilized" nor "human," and when the reality threatening to consume him resembles nothing his mind or soul has ever faced before.

The handful of survivors of the Warsaw ghetto left behind them a world destroyed, one that their lives could not redeem, their memories not revive. The Nazis did their work thoroughly. Nevertheless, like *The Diary of Anne Frank, The Wall* reverberates with its memorable line, its token of hope, its verbal gesture of affirmation: "the only way to answer death is with more life."[13] And abstractly, conceptually, philosophically, it is a noble refrain; but whether Rachel will indeed survive to bear a child in defiance of all these atrocities, we will never know. An even more significant hiatus is the absence of some notion of how such new life "responds" to the multitudinous deaths of the Holocaust. The teasing hint that somehow the future will bring palliatives for such anguish deflects our attention from a harsher truth of the Holocaust, which the play flirts with only to retreat from: that sometimes, many times, the only way to answer your own possible death was with someone else's life. John Hersey, in one crucial episode in the novel at least, does not flinch from this harsher truth: when the baby begins to cry, an underground leader takes it from its mother's arms and smothers it. Doomed people do not behave like men and women in charge of their fate. Lampell was probably correct in assuming that his audience would not tolerate such agony: but *The Wall,* like all art, pays a price for such compromise.

Moral oversimplification is one of the many sins afflicting writing about the Holocaust. We find comfort in schemes of cause and effect: villains destroy; victims submit or resist. We will never understand the behavior of the victims until we gain greater insight into the motives of their murderers. The courageous fighters of the Warsaw ghetto, too few and too late, with scant help from the Polish underground and virtually none from the world outside, knew in advance that they were not choosing life but exerting minimum control over the manner of their death, rescuing a fragment of fate from a seemingly unassailable doom. That doom is personified by their enemies, but in *The Wall* they are mechanized, dehumanized, transformed into robots called merely German Private, German Sergeant, and O. S. Fuehrer. They differ little from the loudspeaker attached to a linden tree, barking orders for assembly and "resettlement." But the Jews were killed by men and women like themselves, not by automatons; one of the play's major faults is its failure to confront the challenge of characterizing those instruments of doom who, through a combination of ruthlessness and manipulation, deprived their victims of moral space to maneuver with dignity.

Regardless of their artistic merit, the plays and films we are examining share one common purpose: they bring us into the presence of human beings searching for a discourse commensurate with their dilemma. In order to recognize that dilemma (the threat of extermination), they must find a language adequate to express it; but in order to find that language, they must first be able to imagine the dilemma. Without such perception, without the words to articulate it to others and make it credible, the individual remains totally vulnerable. Even *with* such "ammunition," after a certain point in time, the Jew had few meaningful alternatives (though many hopeless ones); those left with moral space to maneuver with dignity were the non-Jews, and to his credit Arthur Miller in *Incident at Vichy* (1964) shifts the center of moral responsibility for the situation of the Jews from the victims to the well-intentioned spectator, who begins by dismissing Nazism as "an outburst of vulgarity," but ends by realizing that a more precise definition must move him from language and perception to deeds—a process similar to the one followed by the Warsaw ghetto fighters, though their decision to act could not support anyone's survival.

Except for a gypsy, all the characters in *Incident at Vichy* have been rounded up because they are suspected of being Jews. How is the Jew to protect himself against such danger, psychologically, without supporting his enemy's view of him as a menace to society? He was vulnerable precisely because he was *not* able to see himself in this way: "You begin wishing you'd committed a crime, you know? Something definite," says Lebeau, the painter. "I was walking down the street before, a car pulls up beside me, a man gets out and measures my nose, my ears, my mouth, the next thing I'm sitting in a police station—or whatever the hell this is here—and in the middle of Europe, the highest peak of civilization!" Kafka could impose on life the discontinuities of art, and the courageous reader might follow his vision with impunity; but when life imposed on the Jews Kafkaesque discontinuities like the one Lebeau describes, the confused and terrorized victim could not retreat for insight or relief to the sanctuary of art. Indeed, *Incident at Vichy* may be seen as concisely dramatized dialogues between points of view which question the power of art—on the Holocaust theme—to achieve these ends. Insight and relief are allied to meaning, but when one of the prisoners mouths the cliché that one "should try to think of why things happen. It helps to know the meaning of one's suffering," the painter—recalling that he has been arrested and possibly doomed because of the size of his nose—acidly replies: "After the Romans and the Greeks and the Renaissance, and you know what this means?[14]

When humanistic precedents collapse, the individual loses the security of collective identity; neither family nor group nor profession protects: the prisoners in this play are isolated, alone, searching for private strategies to ensure their release—unaware that the Nazi determination to destroy all Jews has deprived them of choice. The contest is unequal before it begins. If art is an illusion we submit to for greater insight, life—the life depicted in *Incident at Vichy*—is an illusion we submit to from greater ignorance. Habitual ways of thinking become parodies of insight; to rumors about Auschwitz the actor Monceau replies: "Is that really conceivable to you? War is War, but you still have to keep a certain sense of proportion. I mean Germans are still *people*." His naïveté about the equivalence of words to facts breeds futile hopes. Incapable of suspecting that the Germans will identify Jews by ordering them to drop their trousers, he urges his fellow prisoners not to look like victims: "One must create one's own reality in this world. . . . One must show them the face of a man who is right." Unlike the authors of *Anne Frank* and *The Wall*, Miller has the artistic integrity to expose the impotence of such facile rhetoric. Even more irrelevant is the electrician Bayard's conviction that one must not respond personally to the Nazi threat: they may torture the individual, but "they can't torture the future." Only Von Berg, the non-Jew "spectator" who knows that he can secure release by identifying himself, moves toward a clear perception of the future: "What if nothing comes of the facts but endless, endless disaster?"[15]

Miller has a lucid sense of the impotence of the Jewish victims, once they have lost the flexibility of their freedom. They talk of overpowering the solitary guard and fleeing, but a combination of fear, uncertainty, and lack of weapons frustrates this plan, especially after a sympathetic major warns them that armed guards stand outside. They berate themselves for not having been more wary *before* their arrest. But as Von Berg implies, decent men do not possess an imagination for disaster, especially one ending in gas chamber and oven. This is the source of Nazi strength, he argues: they do the inconceivable, and "it paralyzes the rest of us." The consequences for values, for human behavior, are alarming, if not revolutionary: "You must not calculate these people with some nineteenth-century arithmetic of loss and gain. . . . [I]n my opinion, win or lose this war, they have pointed the way to the future. What one used to conceive a human being to be will have no room on this earth." One would like to believe that this is merely the language of fashionable despair, coming from an Austrian aristocrat who can still describe the Nazi menace as "the nobility of the totally vulgar." The doctor Leduc, Prince Von Berg's chief disputant and interlocutor as the others "disappear" into the inner office, refines the Austrian's observation by describing how the Nazis capitalize on their victims' habit of projecting their own reasonable ideas—like the impossibility of extermination—into their enemy's head. "Do you understand?" he pleads. "You cannot wager your life on a purely rational analysis of this situation."[16]

As the wait in this anteroom to deportation and death wears on, façades begin to fall away, and certain truths emerge which, if organized into a system of belief, would indeed transform our vision of what it was like to have been alive in that time—to say nothing of our own. Miller provides insight into the psychology—not necessarily of the Jew—but of the *hunted*, the humiliated, the disenfranchised, the abandoned, the scorned. Lebeau the painter admits to believing in the disaster that threatens, and to exposing himself to danger nonetheless: "you get tired of believing in the truth. You get tired of seeing things clearly." To perceive with the illogic of the Nazis, to endorse, for example, the impossibility of being a Jew in Hitler's Europe, was to accept the erosion of one's own humanity. Unwilling to embrace this course, *unable* to do so in the absence of model precedents, the victim was driven through sheer moral weariness to accept his own mortality, his vulnerability: "one way or the other," says the doctor, "with illusions or without them, exhausted or fresh—we have been trained to die. The Jew and the Gentile both."[17]

But even this seemingly final view is only one in a series of carefully orchestrated positions in *Incident at Vichy:* until the play's last moment, Miller withholds the heroic gesture that we have encountered in *Anne Frank* and *The Wall*, and which even he cannot ultimately resist, though he shrouds it with some ambiguity. Holocaust writing itself serves two masters: a clear intellectual perception of how Nazism shrank the area of dignified choice and reduced the options for human gestures; and the instinct to have victims survive heroically even within these less-than-human alternatives. When the German major (who has been struggling, as a military man, against his own association with the SS inquisitor) asks Leduc whether he would refuse, if he were released and the others kept, the doctor murmurs no; and to the more complex (and deliberately nastier) question of whether he would walk out of the door with a light heart, he can only reply: "I don't know." The confrontation is a *locus classicus* for understanding how the Holocaust undermined what Von Berg would have called the "nineteenth-century arithmetic of loss and gain"—or right and wrong. "One man's death is another man's bread" ran a bitter slogan in the deathcamps, and it represents a far deeper and more painful truth than the principle that in spite of everything, men are really good at heart, or that the only way to answer death is with more life.

In the end, Miller concedes, perception changes little: the non-Jew will live not because he is a better man, but because he is not Jewish; the Jew will die, simply because he is Jewish. Such "logic" fits no moral scheme, generates no satisfactory mode of behavior. Viable attitudes may strengthen the moral will, but only viable acts save the physical self. Miller discovers such an act to end his play—Von Berg gives his pass to Leduc, allowing the Jewish doctor to walk safely past the guard while he himself remains behind to face the wrath of the Nazis—but the gesture simply imposes on a hopeless situation the temporary idealism of self-sacrifice. Von Berg has learned how to share the anguish of the Jew, to cross the terrain separating the complicity of silence from the helplessness of the victim. But as the curtain descends, more men rounded up as suspected Jews appear. One can hardly believe that Leduc will get far with the police in close pursuit, and the routine of identification and deportation is about to begin again.

The motive for Von Berg's gesture is left in doubt: Has he tried to save another man, or his own soul? At least he has acknowledged a vital truth of the Holocaust: to be alive while others perish innocently in the gas chambers is itself a form of complicity. The Jew, as Leduc insists, has learned a grim lesson about the nature of man: "that he is *not* reasonable, that he is full of murder, that his ideals are only the little tax he pays for the right to hate and kill with a clear conscience." *He* has seen into the heart of darkness. Despite his lucidity about the "vulgar" Nazis, Von Berg clings to his ancient idealism, as the only prop that still supports his flagging life: "There are people who would find it easier to die than stain one finger with this murder." One might interpret this dialogue as a confrontation between the voice of a doomed man and one who still believes that he can master his own fate. But the irony of the moment is Von Berg's slow realization that he can restore the challenge of fate to Leduc only by surrendering his own, and submitting to the doom that the Jew now hopes to escape. One would feel more elated about this tribute to the human spirit were it not surrounded by the disappearance of the other Jews and the arrival of new victims who will have no one to "save" them. Von Berg has repudiated the idea of man's right to hate and kill with a clear conscience. A world without ideals, as he said earlier, would be intolerable. But how does one measure his private deed of generosity against the slaughter of millions? Does it invalidate Leduc's melancholy charge, only too familiar to survivors of the deathcamps, that "each man has his Jew . . . the man whose death leaves you relieved that you are not him, despite your decency"?[18] Miller at least pays homage to the ambiguity of efforts to redress the imbalance between justice and suffering imposed by the Nazi atrocities. But the magnitude of the sorrow and loss dwarfs the deed, however noble, of one man for one man; *Incident at Vichy* illuminates the difficulty, perhaps the impossibility, of affirming the tragic dignity of the individual man, when it has been soiled by the ashes of anonymous millions.

If *Incident at Vichy* exposes the dilemma of measuring the private deed of generosity against the slaughter of millions, Abby Mann's script for the Stanley Kramer production of *Judgment at Nuremberg* (1961) raises the issue of measuring the public act of justice against that same slaughter. If Miller's play complicates the problem of redressing the imbalance between justice and such suffering, Kramer's film simplifies the question of establishing a connection between mass atrocity and individual responsibility. The force of idealism in *Judg-*

ment at Nuremberg resides not in the faith of a young girl, or a beleaguered ghetto fighter, or an Austrian aristocrat, but in a principle: that in a court of law, with only relevant testimony allowed, where defendants have counsel who may cross-examine witnesses, a just verdict against an unjust act somehow satisfies the conscience even though political expediency—in this instance, the Berlin Airlift and its consequences for German–American relations—may afterward undermine its durability. Just before the final fadeout, the audience is treated to the unsavory irony that by July 14, 1949, after the last of the so-called second Nuremberg trials, of "ninety-nine sentenced to prison terms, not one is still serving his sentence."

I suppose that for those who still see the Holocaust as a situation of violated justice, this announcement is as exasperating and offensive as the news in 1980 that 200 alleged Nazi war criminals are still living safely in the United States. But both truths are accompanied by an overwhelming sense of futility, especially to those who have already understood that the logic of law can never make sense of the illogic of extermination. The film tries to do exactly the reverse; how else are we to interpret its final episode, with its unqualified assumption that the perversion of justice necessarily leads to genocide? The trial over, the guilty verdicts in, the life sentences proclaimed, the chief judge, Dan Haywood (Spencer Tracy), confronts the chief defendant, former Nazi Minister of Justice Ernst Janning (Burt Lancaster), who "doomed" himself in court by honorably announcing that "if there is to be any salvation for Germany those of us who know our guilt must admit it no matter what the cost in pain and humiliation." The two representatives of justice, now and then, face each other, and Janning says to Haywood: "I want to hear from a man like you. A man who has heard what happened. I want to hear—not that he forgives—but that he understands." The kind of insight (if not charity) demanded by this question, the film does not know how to dramatize. Whether because of a failure of nerve, of dramatic sense, or of artistic imagination, the author and director have not solved the problem of exploring this difficult psychological issue in their courtroom drama. Judge Haywood gropes for a response in a scene direction in the script, but this is conveyed to the audience only through the inarticulate furrows of Spencer Tracy's brow: "understand, Haywood says to himself. I understand the pressures you were under. . . . But how can I understand the deaths of millions of men, women, and children in gas ovens [a common technical blunder fusing and confusing gas chambers with crematorium ovens], Herr Janning?" A shrewd interpreter of furrowed brows, Janning replies in stereotypical and (as we shall see) self-contradictory justification: "I did not know it would come to that.

You must believe it. You must believe it." The judge's rejoinder is the closing line of the film: "Herr Janning. It came to that the first time you sentenced a man to death you knew to be innocent."[19] A melodramatic riposte, precise and compact: But are we expected to be content with this explanation of the murder of 6 million Jews? When decent men, through tacit or active consent, from personal gain, weak will, or false patriotism, authorize (as in this film) the sterilization of a Communist's son or the execution of an elderly Jew on the trumped-up charge of cohabiting with an Aryan, they participate in a totalitarian manipulation of the law against certain individuals—but do they lay the groundwork for the annihilation of a people?

That annihilation is central to the historical truth behind the film, if not to its cinematic logic: no serious treatment of the Holocaust can avoid it. But *Judgment at Nuremberg* introduces evidence about Buchenwald and Belsen in an almost gratuitous fashion, though the impact of these actual army films on cast and audience is unquestionable; indeed, instead of establishing continuity between the horrifying events and the deeds of the defendants, the images of atrocity on a screen within a screen create a psychological distance: What system of justice can render homage to the mounds of corpses being shoved into a mass grave by a giant British bulldozer? The real event of atrocity and the subsequent film using it as a background do not cohere in a satisfying image of deed and consequence, uniting the twisted bodies of the victims with the acts of the four men on trial. At the end, Janning protests, "I did not know it would come to that." But earlier, during his moment of "confession," he is more precise: "Maybe we didn't know the details. But if we didn't know, it was because we didn't want to know." The space between these two statements, between denial and equivocation, forms a psychological desert that the film doesn't begin to explore. Yet that wasteland is exactly the area whose detailed examination would make a film like *Judgment at Nuremberg* worthy of our attention. Three of the four German judges on trial are puppets in the dock, masks of contempt, indifference, or apprehension, but never human beings. The unrepentant Nazi, the timid collaborator, and the perennial self-justifier are familiar stereotypical figures whose presence in the film offer no insight whatsoever into the nature of the genocidal impulse. And Ernst Janning doesn't help us by denouncing himself as worse than any of them "because he knew what they were and went along with them." Nor does his self-portrait, shaded with aristocratic disdain—he "made his life . . . excrement because he walked with them"[20]—suggest any of the moral chiaroscuro that might fall from the master hand of a Rembrandt.

Janning's declaration of guilt is not enough, psychologically or artistically: not enough for himself, because he still has not penetrated his motives, and not enough for us, because we still do not know how such a decent man came to lend his judicial prestige to the Nazi cause.

Admission of guilt neither restores decency—the lingering horrors of Belsen ensure this in the film—nor explains its perversion, any more than Albert Speer's acknowledgment (at the primary Nuremberg trial and in his memoirs) of responsibility for Nazi crimes he claims to have known nothing about restores our faith in him or explains the tangled motives that led him to embrace the aspirations of the Third Reich. Nor does the eloquent speech of Rolfe the defense attorney, ably portrayed by Maximilian Schell, take us any deeper into the heart of Holocaust darkness. By expanding the perimeters of guilt—"The whole world is as responsible for Hitler as is Germany"[21]—Rolfe seeks to mitigate blame for his client; but under the burden of six million innocent victims, the structure of this argument collapses too. Although the immediate thrust of the film is to leave the audience feeling the irony of the conflict between political expediency and absolute justice, the perhaps unintended momentum of its subterranean current—reinforced by the impact of the documentary views of the camps—is to oppress the audience with the irrelevance of absolute justice to the collective crime of mass murder. In the aftermath of such "indecent" dying, the courtroom as mis-en-scène for a comprehension of its enormity seems pitifully inadequate. Like most Hollywood efforts, this admittedly serious film scants the more complex interior landscape of the mind, where Judge, Victim, and Accused surrender the clear outlines of their identity and wrestle with a reality that subverts the very meaning of decency. *Judgment at Nuremberg,* with its concluding irony that by July 1949 none of the ninety-nine sentenced to prison at the last of the second series of trials was still behind bars, renders a judgment *of* Nuremberg, of Nazism, and of the contemporary world that it probably never wished to impose: that uncorrupted justice, the highest expression of law, order, morality, and civilization, is only a charade in the presence of atrocities literally embodied by the mounds of twisted corpses in mass graves at Belsen. The film uses the Holocaust theme only to misuse it; it focuses momentarily on the original horror, then shifts our awareness (thereby once more warming our hearts) to the admirable probity of an amiable American judge. If the irony of history, allying Germany with one former enemy against another former one, has the last word, what happens to the responsibility of art, which must avoid the trivialization of that horror despite history's insensitivity? This is the real weakness of *Judgment at Nuremberg,* which mirrors that insensitivity while pretending that the temporary triumph of justice over expediency—German or American—can make a difference.

"To produce a mighty book," Melville says in *Moby Dick,* "you must choose a mighty theme." Perhaps modesty kept him from adding that a mighty talent was also necessary. Few contemporary themes present "mightier" challenges to the imagination than the Holocaust, the implacable, mesmerizing White Whale of our time; but when American television committed millions to produce Gerald Green's documentary drama, *Holocaust* (1978), those who provided the money did not suspect that they had invested in a dolphin, though having been promised a whale. Had the makers of this mammoth enterprise been more attuned to sacred Scripture, they might have been forewarned of the danger of trying to draw out leviathan with a hook.

When a famous survivor of Auschwitz, Elie Wiesel, protested in print that, whatever its intentions, *Holocaust* "transforms an ontological event into soap-opera,"[22] Green defended himself by citing the contrary views of distinguished critics like John Rich, Tom Shales, and Harriet Van Horne. To justify himself against the charge of large-scale mingling of fiction with historical fact, he cited the earlier example of *War and Peace,* compounding his sponsor's mistake of confusing dolphins with whales. The real issue was lucidly raised by *Time* writer Lance Morrow: "one senses something wrong with the television effort when one realizes that two or three black-and-white concentration-camp still photographs displayed by Dorf [fictional SS officer]—the stacked, starved bodies—are more powerful and heartbreaking than two or three hours of dramatization." Morrow goes on to complain that the "last 15 minutes of Vittorio De Sica's *The Garden of the Finzi-Continis,* in which Italian Jews are rounded up to be taken to the camps, is more wrenching than all the hours of *Holocaust*"[23]—a judgment with which I happen to agree. At least two crucial issues are raised here, both of which need to be addressed before anyone can respond intelligently to the television drama *Holocaust.*

The first is the question of whether that "ontological event" *can* be transformed into a form of artistic experience carrying it beyond its historical moment and making it accessible in all its complexity to those who have not directly experienced it. Can representation rival the immediacy of contemporaneous photographs, diaries dug up after the war in the ashes surrounding Auschwitz, or even survivor accounts—unromantic emotions recollected in disquietude? Nothing can rival the chaotic masses of undifferentiated corpses, the token of men's doom in the Holocaust; but I would suggest that only art can lead the uninitiated imagination from the familiar realm of man's fate to the icy atmosphere of the deathcamps, where collective doom replaced the private will. Holocaust art is transitional art, a balloon, as it were, straining to break free from its inspiring reality but always moored by a single stubborn strand to the ontological event that gave it birth—the extermination of millions of innocent human beings. It is a necessary art, ever more necessary as that event recedes in time and new generations struggle to comprehend why a civilized country in the midst of the twentieth century coolly decided to murder all of Europe's Jews. The documents themselves do not answer this vital question for us.

Does television's *Holocaust?* Many who celebrate the film argue that for the first time, at least in America, viewers will get a sense of what it was like to suffer being Jewish during the Nazi period. Nothing before on American stage or screen had faced so "fearlessly" the fact of genocide, and the process which led up to it. The uninitiated imagination is offered Kristallnacht, the euthanasia program, the Warsaw ghetto, Buchenwald, Theresienstadt, Auschwitz, Sobibor, and Babi Yar—to say nothing of Eichmann, Heydrich, Ohlendorf, Biberstein, Blobel, and several others in the hierarchy of Nazi executioners. One sees Jews being beaten, starved, tortured, marching primly into the gas chambers of Auschwitz, tumbling docilely into the murder pits at Babi Yar. One hears Erik Dorf and his historical counterparts speaking of Jews as if they were vermin, superfluous equipment, detritus on the shores of Europe. Why is it then that even Kurtz's Intended might have watched this spectacle without feeling that it was too dark—too dark altogether? Why is it that nothing in the drama equals or even approaches the unmitigated horror of the actual films of Nazi executions which Dorf shows his superiors? One might argue that art never matches history; but in the case of *Holocaust*, it is more valid to conclude that talent has not matched intention. The failure of *Holocaust* is a failure of imagination. The vision which plunges us into the lower abysses of atrocity is not there. We do not know what it was like, in the Warsaw ghetto and elsewhere, to have been reduced to eating dogs, cats, horses, insects, and even, in rare unpublicized instances, human flesh. We do not know what the human being suffered during days and nights in sealed boxcars, starving, confused, desperate, sharing one's crowded space with frozen corpses. We do not know of the endless roll calls in Auschwitz, often in subfreezing temperature, when men and women simply collapsed and died from exhaustion. We have abundant examples of husbands and wives clinging together in adverse conditions, but we never glimpse—as I mentioned earlier—mothers abandoning children or fathers and sons throttling each other for a piece of bread. We see well-groomed and sanitized men and women filing into the gas chamber, but what does this convey of the terror and despair that overwhelmed millions of victims as they recognized the final moment of their degradation and their powerlessness to respond? Perhaps art will never be able to duplicate the absolute horror of such atrocities: but if it cannot re-create at least a limited authentic image of that horror—and *Holocaust* does not—then audiences will remain as deceived about the *worst* as young Anne Frank's lingering words on the essential goodness of human nature deceive us about the *best.*

A second crucial test of all Holocaust art is the question of insight. Having viewed *Holocaust,* do we have any *fresh* insights into the Nazi mind, the victim, the spectator? The drama chooses the least arduous creative path, tamely following chronological sequence, though the disruption of time-sense (and place-sense) was one of the chief features of the deathcamp experience: tomorrow vanished, and the past became a dim, nostalgic echo. It adopts the safe strategy of externalizing the event. A doctor sends Anna Weiss to Hadamar, notorious euthanasia center in prewar Germany, but neither he, nor her sister-in-law (who consents to let her go), nor the nurses who receive her, nor the group of puppetlike figures who join her in the walk to their death, reveal any vivid anguish, any searing conflict, any terror, pain, or even dislocation of the moral center of their being. If extermination *is* such a simple matter of banality (and I do not believe it), it is hardly worth writing about. Even more impoverished (in terms of fresh insight) is the spectacle of Dr. Weiss and his friend Lowy lamely walking to their doom in the gas chambers of Auschwitz. What do men think or feel at that moment? In this critical instant of an "ontological event" that remains an abiding trauma to the modern imagination, the doctor mutters something about his never having time now to remove his friend's diseased gall bladder. Do we have a right to expect more, before we bestow the laurel of insight on *Holocaust?*

Could Gerald Green and his producers have done better? Consider this brief portrait of two human beings, about to die in the gas chambers of Auschwitz, recorded by Salmen Lewental in a diary exhumed from the ashes after the war:

> A mother was sitting with her daughter, they both spoke in Polish. She sat helplessly, spoke so softly that she could hardly be heard. She was clasping the head of her daughter with her hands and hugging her tightly. [She spoke]: "In an hour we shall die. What a tragedy. My dearest, my last hope will die with you." She sat . . . immersed in thought, with wide-open, dimmed eyes. . . . After some minutes she came to and continued to speak. "On account of you my pain is so great that I am dying when I think of it." She let down her stiff

> arms and her daughter's head sank down upon her mother's knees. A shiver passed through the body of the young girl, she called desperately "Mama!" And she spoke no more, those were her last words.[24]

Perhaps "last words" like these are not dramatic; certainly, they are not commercial; undoubtedly, they are not American. But they are authentic, and they are what the Holocaust was all about. The upbeat ending of *Holocaust*, minimizing the negative impact of all that has gone before, typifies the absence of insight and the externalization of horror that makes the entire production meretricious in its confrontation with disaster: wormwood and gall are mollified by aromatic spices from the Orient. To leave an audience of millions with an image like the one of mother and daughter bereft of hope, of life, of speech would have been too dark—too dark altogether. The American theater and screen, the American mind itself, is not yet ready to end in such silence. The heroic gesture still seizes us with its glamor, tempering the doom of men and women who have lost control of their fate. Salmen Lewental has recorded an epigraph for all writing about victims of the Holocaust, whether as art or history—and it may serve as epitaph too: "On account of you my pain is so great that I am dying when I think of it." The memory of 6 million dead echoes as a symphony of pain: in that denial of final triumph lies our acceptance and understanding of the Holocaust experience.

1983

14

What More Can Be Said About the Holocaust?

can hardly remember a Holocaust conference I've attended during the past decade where someone hasn't echoed George Santayana's solemn platitude that those who ignore the past are doomed to repeat it. The so-called ethnic cleansing in Yugoslavia is only one of a dozen episodes in recent years to prove that Santayana's maxim is nothing more than a piece of rhetorical excess—yet we continue to use it as if it played a vital role in defining the meaning of atrocity in our time. Indeed, it could be argued about the violence in Bosnia that the contending forces not only have *not* ignored the past of the Holocaust, but have paid careful attention *to* it in order to learn more about how to dehumanize their enemy in the name of some purifying ideology.

In other words, understanding the past does not necessarily help us to avoid *or* to predict undesirable futures. What Santayana perhaps should have said is that those who ignore the past are doomed to remain ignorant not only of that past, but especially of how the choices of individual human beings sometimes interact with the chances of political reality to create unprecedented historical occasions—whether the creation of a new civilization or the destruction of an entire people. More specifically, we still can't fathom how the appearance of a man called Adolf Hitler ended with the disappearance of most of European Jewry, or how thousands of men and women who ordinarily would not even consider injuring another person found themselves willingly participating in the event we name the Holocaust. Furthermore, we still can't grasp the inner workings of what I call the "aversion-to-crisis faculty," which kept both potential victims and the nations that might have given them refuge from acknowledging the fatal threat to European Jewry that was expanding almost daily. Fifty years from now, our descendants will be asking the same questions about *our* paralysis before the AIDS epidemic, which, according to some estimates, in a decade may consume seven or eight times as many victims as did the Holocaust.

My point is this: every future generation will have to be educated anew in how to face the historical period we call the Holocaust. This must be done not through abstract formulas like "the murder of 6 million," but in graphic detail, so that the destruction of an entire people and its culture—what was done, how it was done, and by whom—makes an indelible and subversive impression on their moral, political, philosophical, and psychological assumptions about individual behavior, the nature of reality, and the process of history. The implications of the Holocaust are so bleak that we continue to wrestle with the desperate issue of how best to represent it. That problem still needs to be solved. Literature, history, testimony, commentary, theological speculation—many avenues exist for entering its vestibule, but no two approaches offer identical visions to those who cross the threshold into the landscape of the Holocaust itself.

Years ago, Jakov Lind wrote a novel about the Third Reich called *Landscape in Concrete*. Few people read it then, and no one reads it today, but its central image of a panorama of stone rubble as a major legacy of World War II is not unfamiliar to those who remember the site of the Warsaw ghetto after its destruction. Indeed, the city of Warsaw, to say nothing of major metropolises like Berlin and Hamburg in Germany itself, extend the range and relevance of that image. Lind's landscape of devastation, which in addition to its other victims consumed 6 million Jews and the communities that nurtured them, is a lifeless terrain. Yet today life flourishes in most of these former localities of death, and little remains, except for a few museums, monuments, and partly preserved camps, to remind the visitor of what has been lost.

For this reason, I think the challenge facing future Holocaust specialists is to reverse history and progress and find a way of restoring to the imagination of coming generations the depth and scope of the catastrophe. We can do this in a number of ways. I have no doubt that recently opened archives in Eastern Europe and the Soviet Union will provide all kinds of fresh information on the fate of Jewish populations in those regions. The history of no era is ever complete, and the Holocaust in particular leaves many unanswered questions to be investigated. Not so long ago, some newly discovered Paraguayan police files insisted that Martin Bormann, Hitler's civilian chief of staff, as it were, died in that country on February 15, 1959, and was buried not far from the capital, Ascunción, two days later. Many have long been skeptical of the accepted view that Bormann was killed or committed suicide in Berlin near the end of the war, but until now little evidence to dispute that belief had surfaced. Yet the longtime head of the Nazi Records Center in Ludwigsburg, Germany, a very knowledgeable official, called the report "complete nonsense" and insisted that Bormann's skull had been found in Berlin, where he most certainly died in 1945. The argument will no doubt be settled, and one claim or the other certified as correct. This of course is only a detail, but it calls attention to the need to reexamine regularly many of the accepted conclusions about the Holocaust experience, lest we grow complacent and embrace final answers when we should still be pursuing elusive questions.

The details of the Holocaust are not the only area of concern. Attitudes toward it are equally and perhaps more important. On this subject, much needs to be *un*said, while even more remains to be said. As happens with other historical events, a Bruno Bettelheim or a Viktor Frankl or a Robert Jay Lifton or even a Raul Hilberg establishes a way of perceiving a particular realm of the Holocaust, and it becomes the "authorized" view that future students of the Holocaust uncritically absorb. Bettelheim's early charge that the least successful camp inmates regressed to passive and childlike behavior instead of choosing defiant and assertive postures was echoed by Hilberg in the opening section of his seminal work *The Destruction of the European Jews*, where he declares that compliance was an essential feature in the conduct of many Jews of Eastern Europe. These approaches, in turn, helped to inspire the "blame-the-victim" formula that still haunts us today, in the cliché that the Jews went like sheep to the slaughter. Frankl almost single-handedly invented the idea of spiritual resistance in *Man's Search for Meaning*, a work that appeared in its original German edition right after the war and still provides millions of readers with a solace that blurs the true painful nature of the deathcamp ordeal. So instinctive is the appeal of Frankl's vision, so essential to protect us from the grim visage of atrocity, that fifty years after the event we still have before us the task of educating audiences in the experience of confrontation.

Certain pivotal works on the Holocaust (or works accepted as pivotal) shape habits of mind that are difficult to dislodge, and I believe one of the main tasks of future historians and commentators is to identify such works and to subject them to a fresh scrutiny, so we can see whether the habits of mind they promote really provide greater insight into the issues we still need to investigate.

Two issues—and, of course, there are others—that are in their infancy as far as exhaustive analysis and interpretation is concerned are the circumstances of the victims in the camps and the motives of the men and women who organized, facilitated, and were directly instrumental in their deaths. The former is easier to address than the latter, since so much more trustworthy material is available, though for reasons that perplex me, historians seem reluctant to explore the oral testimonies of surviving victims while turning with fascination to the oral testimony of the perpetrators. Lifton's massive study of Nazi doctors is only one example of a researcher's willingness to listen patiently to the voices of Holocaust collaborators and to study their motives. The category of "doubling" that Lifton proposes in order to explain how professionally trained physicians managed to separate their gruesome duties in the camps from their "normal" life outside suggests a process of moral differentiation that perhaps satisfies the analyst-researcher more than the actual facts. The widespread absence of remorse among the accused in postwar trials indicates that we may need to discard familiar categories like doubling and to accept the possibility of a regimen of behavior that simply dismisses conscience as an operative moral factor. The notion of the power to kill, or to authorize the killing of others, as a personally fulfilling activity is not appealing to our civilized sensibilities; even more threatening is the idea that this is not necessarily a pathological condition, but an expression of impulses as native to our selves as love and compassion. I think we will not begin to understand how the Holocaust could have happened until we abandon simplified moral and psychological categories like doubling and conscience and regard the mind of the perpetrators as another landscape in concrete, alive but lacking the vital signs we are accustomed to seeing there.

As for the victims, Israeli novelist Aharon Appelfeld has outlined the charge to future commentators. Although he is speaking of literature, his remarks need not be confined to the practitioners of that discipline:

> By its very nature, when it comes to describing reality, art always demands a certain intensification, for many and for various reasons. However, that is not the case with the Holocaust. Everything in it already seems so thoroughly unreal, as if it no longer belongs to the experience of our generation, but to mythology. Thence comes the need to bring it down to the human realm. This is not a mechanical problem, but an essential one. When I say "to bring it down," I do not mean to simplify, to attenuate, or to sweeten the horror, but to attempt to make the events speak through the individual and in his language, to rescue the suffering from huge numbers, from dreadful anonymity, and to restore the person's given and family name, to give the tortured person back his human form, which was snatched away from him.[1]

Sweetening the horror of the Holocaust has become the pastime of so many students of the subject that they are too numerous to list. Readers return to these sources like bees to their hive, because the honeyed vision they find there seems so savory to their moral taste buds. Even a critic as astute as Terrence Des Pres, who certainly was no stranger to the dark side of the Holocaust, could not in his influential work *The Survivor* resist arguing that survivors share "a past identical for everyone who came through the common catastrophe." Anyone conversant with the thousands of separate testimonies of survivors knows what an *un*common catastrophe it was for each of them, no two of whom feel that their pasts were identical. By creating a collective identity called "the survivor" in order to celebrate a talent for life, "which enables men and women to act spontaneously and correctly during times of protracted stress and danger,"[2] Des Pres fortifies a myth instead of exposing a truth. His is only one voice among many to ignore Appelfeld's plea to "rescue the suffering from huge numbers, from dreadful anonymity, and to restore the person's given and family name, to give the tortured person back his human form, which was snatched away from him." Because that human form often resembles an austere etching by Leonard Baskin rather than a lavish portrait by Whistler, we too often turn away from it and prefer to celebrate the idea that there was a "correct" way to behave for people facing the threat of starvation and the gas chamber. One of the Holocaust mysteries remaining to be explored is why so many sensitive and educated minds still flinch from the awful realities of mass murder and persist in weaving elaborate fantasies about the dignity of dying under or living through such miserable circumstances.

It seems that two forces are at work in Holocaust commentary, and much more needs to be said about the origins of the conflict—both at home, I might add, and abroad. Sometimes the motives are well intentioned, sometimes not, but what we appear to have is, on the one hand, a historical consciousness determined to distort or at least alleviate the harshest truths of the Holocaust, and, on the other, a historical consciousness resolved to confront its implications wherever they may lead. Israeli-German historian Frank Stern has spoken recently of "a clash of antagonistic memories," and future Holocaust commentary may very well be forced to focus disproportionately on this issue. He describes the emergence of a "myth of the new German innocence":

> [A]n ambiguous form of German historical consciousness is developing which promotes an emotional and intellectual distance from the Holocaust for the sake of a new German national identity and, ultimately, the creation of a new German national state. The Germans

> are going back to the future, and are doing so on the basis of a highly selected past. Since the impact of the Holocaust cannot be negated, its meaning has to be changed and integrated in an almost harmless way in the new German national thinking.[3]

The need to make the Holocaust appear more harmless than it was has many roots, and hence many branches, leaves, and blossoms. Its efforts to sweeten the bitter fruits of mass murder will have to be monitored for decades, and perhaps generations, if we are to prevent what happened from slipping into a vague limbo of forgetfulness, a footnote to contemporary history instead of the central historical moment of our time, and perhaps of all time.

Stern opposes a Jewish view of the past against a German view of the past, but I would prefer to speak of a more universal division between our human and our inhuman past. The essential *inhumanity* of this event is the gadfly that vexes our perception: Germans do not distance themselves from World War I or the Franco-Prussian War. Whether the cause is national guilt or private conscience or religious conviction or the desire to preserve intact a system of coherent moral values, they all betray a paralyzing dismay at what the victims have suffered and the agents have done. Half a century old, the Holocaust still mocks the idea of civilization and threatens our sense of ourselves as spiritual creatures. Its undiminished impact on modern memory leaves wide open the unsettled and unsettling question of why this should be so.

1993

Notes

Introduction

1. Quoted in David Remnick, "Profile: The Exile Returns;" *New Yorker*, 14 February 1994, p. 73.

2. Harold Brodkey, "Dying: An Update," *New Yorker* 7 February 1994, p. 73.

Chapter 1

1. Abraham Lewin, *A Cup of Tears: A Diary of the Warsaw Ghetto*, ed. Antony Polonsky, trans. Christopher Hutton (Oxford: Basil Blackwell, 1988), p. 228.

2. Jean-François Lyotard, *Heidegger and "the jews,"* trans. Andreas Michel and Mark S. Roberts (Minneapolis: University of Minnesota Press, 1990), p. xxii.

3. Ibid., p. 10.

4. Ibid., p. 12.

5. Ibid., p. 16.

6. Ibid., p. 23.

7. Ibid., pp. 26, 33.

8. Jean-François Lyotard, *The Differend*, trans. Georges van den Abbele (Minneapolis: University of Minnesota Press, 1988), p. 109.

9. Lyotard, *Heidegger and "the jews"* p. 47.

10. Lyotard, *The Differend*, p. 100.

11. Ibid.

12. Lyotard, *Heidegger and "the jews,"* p. 79.

13. Fortunoff Video Archive for Holocaust Testimonies, Yale University, tape T-938: testimony of George S.

Chapter 2

1. Victor Frankl, *Man's Search for Meaning*, rev. and updated ed. (New York: Pocket Books, 1984), p. 88.

2. Fortunoff Video Archive for Holocaust Testimonies, Yale University, tape T-738: testimony of Abraham P.

Chapter 3

1. Quoted in Hannah Krall, *Shielding the Flame: An Intimate Conversation with Dr. Marek Edelman, the Last Surviving Leader of the Warsaw Ghetto Uprising,* trans. Joanna Stasinska and Lawrence Weschler (New York: Holt, 1986), p. 9.

2. Yisrael Gutman, *The Jews of Warsaw, 1939–1943: Ghetto, Underground, Revolt,* trans. Ina Friedman (Bloomington: Indiana University Press, 1989), p. 207.

3. Quoted in Krall, *Shielding the Flame,* p. 9.

4. Ibid., p. 6.

5. Abraham Lewin, *A Cup of Tears: A Diary of the Warsaw Ghetto,* ed. Antony Polonsky, trans. Christopher Hutton (Oxford: Basil Blackwell, 1988), p. 243.

6. Emmanuel Ringelblum, *Notes from the Warsaw Ghetto: The Journal of Emmanuel Ringelblum,* ed. and trans. Jacob Sloan (New York: Schocken, 1978), p. 310.

7. Lewin, *Cup of Tears,* pp. 153–154.

8. Ibid., pp. 154, 156.

9. Ibid., p. 161.

10. Ibid., p. 73.

11. Ibid.

12. Adina Blady Szwajger, *I Remember Nothing More: The Warsaw Children's Hospital and the Jewish Resistance,* trans. Tasja Darowska and Danusia Stak (New York: Pantheon, 1990), p. 571.

13. Ibid., p. 58.

14. See Charlotte Delbo, *Days and Memory,* trans. Rosette Lamont (Marlboro, Vt.: Marlboro Press, 1990), pp. 1–4.

15. Quoted in Krall, *Shielding the Flame,* p. 92.

16. Szwajger, *I Remember Nothing More,* p. 166.

17. Gutman, *Jews of Warsaw,* p. 390.

18. Quoted in Claude Lanzmann, *Shoah: An Oral History of the Holocaust* (New York: Pantheon, 1985), pp. 197–198.

19. Ibid., p. 198.

20. Ibid., p. 196.

Chapter 4

1. "A Girl's Diary," in *Lodz Ghetto: Inside a Community Under Siege,* comp. and ed. Alan Adelson and Robert Lapides (New York: Viking, 1989), p. 240.

2. Ibid., pp. 242–243.

3. Ibid., p. 245.

4. David Sierakowiak, "Diary," in *Lodz Ghetto,* ed. Adelson and Lapides, p. 143.

5. *The Warsaw Diary of Adam Czerniaków: Prelude to Doom*, ed. Raul Hilberg, Stanislaw Staron, and Josef Kermisz, trans. Stanislaw Staron and staff of Yad Vashem (New York: Stein and Day, 1979), p. 237.

6. Avraham Tory, *Surviving the Holocaust: The Kovno Ghetto Diary*, ed. Martin Glibert, trans. Jerzy Michalowicz (Cambridge, Mass.: Harvard University Press, 1990), p. 51.

7. Chaim Rumkowski, "Give Me Your Children," in *Lodz Ghetto*, ed. Adelson and Lapidus, pp. 328, 331.

8. Ibid., p. 329.

9. Ibid., pp. 329, 330.

10. Tadeusz Borowski, *This Way for the Gas, Ladies and Gentlemen and Other Stories*, trans. Barbara Vedder (New York: Viking, 1967), pp. 101–102.

11. "A Father's Lament," in *Lodz Ghetto*, ed. Adelson and Lapidus, p. 348.

Chapter 5

1. Josef Bor, *The Terezín Requiem*, trans. Edith Pargeter (New York: Knopf, 1963), passim. For a corrective commentary on Bor's semifictionalized version, see Joza Karas, *Music in Terezín: 1941–1945* (New York: Beaufort/Pendragon, 1985), pp. 139–141.

2. Alfred Kantor, "Introduction," *The Book of Alfred Kantor* (New York: McGraw-Hill, 1971).

3. Ibid.

4. Jean Améry, *Jenseits von Schuld und Sühne* (Munich: Szczesny, 1966), p. 33. [My translation]

5. Janusz Korczak, *Ghetto Diary* ed. Aaron Zeitlin (New York: Holocaust Library, 1978), p. 186, n. 7. Korczak could not have been aware of the ironic circumstance that many Jews deported to Auschwitz (rather than Treblinka) were greeted on the ramp by Dr. Josef Mengele, later known by some of his victims as "the angel of death."

6. Yankel Wiernik, "A Year in Treblinka," in *The Death Camp Treblinka*, ed. Alexander Donat (New York: Holocaust Library, 1979), p. 163.

7. Haas's painting *Expecting the Worst* is reproduced in Gerald Green, *The Artists of Terezín* (New York: Hawthorn, 1969), p. 106.

8. Fritta's painting *Quarters of the Aged* is reproduced in ibid., p. 109.

9. Only the first volume, *None of Us Will Return*, trans. John Githens (New York: Grove Press, 1968), has been translated into English. The second volume is *Une Connaissance inutile* (Paris: Éditions de Minuit, 1970); the third is *Mesure de nos jours* (Paris: Éditions de Minuit, 1971). A related volume, posthumously published, has appeared recently: *Days and Memory*, trans. Rosette Lamont (Marlboro, Vt.: Marlboro Press, 1990). For a more positive view of the role of cultural activity in Delbo's camp experience, see Ellen Fine, "Literature as Resistance: Survival in the Camps," *Holocaust and Genocide Studies: An International Journal* 1, no. 1 (1986): 79–89.

10. Charlotte Delbo, "Phantoms, My Companions," trans. Rosette C. Lamont, *Massachusetts Review* 12, no. 1 (Winter 1971): 23, 24, 28–29, 30.

11. Delbo, *Une Connaissance inutile*, p. 90. [My translation]

12. Abraham Sutzkever, "Burnt Pearls," in *Burnt Pearls: Ghetto Poems of Abraham Sutzkever*, trans. Seymour Mayne (Oakville, Ont.: Mosaic Press, 1981), p. 38.

Chapter 6

1. Günther Schwarberg, *The Murders at Bullenhuser Damm*, trans. Erna Baber Rosenfeld with Alvin Rosenfeld (Bloomington: Indiana University Press, 1984).

2. *An Interrupted Life: The Diaries of Etty Hillesum, 1941–1943*, trans. Arno Pomerans (New York: Pantheon, 1983).

3. And one does hear more from her. Additional letters from the Dutch transit camp were published in English three years later as *Letters from Westerbork* (New York: Pantheon, 1986). Survivors remember her "shining personality" there, and she herself wrote that she wanted to become the "thinking heart of the barracks." Consumed by a desire to help and solace, she gave less space in these letters to her own spiritual ambitions, choosing instead to ease the needs of future deportees. Hence she became more sensitive to the "web of sorrow" in which they were entangled and left some memorable descriptions of the grim pilgrimage that was to end in the death of much of Dutch Jewry, including her and her family.

Chapter 7

1. Norma Rosen, "The Second Life of Holocaust Imagery," *Witness* 1, no. 1 (Spring 1987): 14.

2. A. Anatoli (Kuznetsov), *Babi Yar: A Document in the Form of a Novel*, trans. David Floyd (New York: Farrar, Straus & Giroux, 1970), p. 98.

3. D. M. Thomas, *The White Hotel* (New York: Pocket Books, 1981), p. 287.

4. Saul Bellow, *Mr. Sammler's Planet* (New York: Viking, 1970), pp. 289–290. [Subsequent references are in the text]

5. William Styron, *Sophie's Choice* (New York: Random House, 1979), p. 483. [Subsequent references are in the text]

Chapter 8

1. Bernd Naumann, *Auschwitz: A Report on the Proceedings Against Robert Karl Ludwig Mulka and Others Before the Court at Frankfurt*, trans. Jean Steinberg (London: Pall Mall Press, 1966), pp. 146, 147.

2. Quoted in Viktor Frankl, *Man's Search for Meaning*, rev. and updated ed. (New York: Pocket Books, 1984), p. 87.

3. Ibid., p. 88.

4. Naumann, *Auschwitz*, p. 217.

5. Jean Améry, *At the Mind's Limits: Contemplations by a Survivor on Auschwitz and Its Realities,* trans. Sidney Rosenfeld and Stella P. Rosenfeld (New York: Schocken, 1986), p. 16.

6. Frankl, *Man's Search for Meaning,* p. 89.

7. Améry, *At the Mind's Limits,* p. 17.

8. Ibid., pp. 16–17.

9. Ibid., p. 19.

10. Ibid., p. 17.

11. Primo Levi, *The Reawakening: A Liberated Prisoner's Long March Home Through East Europe,* trans. Stuart Woolf (Boston: Little, Brown, 1965), p. 12. Although the book appeared in Italy in 1963, Levi says he finished writing it in 1947.

12. Ibid., pp. 12–13.

13. Ibid., p. 13.

14. Rolf Hochhuth, *The Deputy,* trans. Richard Winston and Clara Winston (New York: Grove Press, 1964), p. 222.

15. Ibid., p. 223.

16. Ibid., p. 249.

17. Peter Weiss, *The Investigation,* trans. Jon Swan and Ulu Grosbard (New York: Atheneum, 1966), pp. 108, 107–108.

18. Tadeusz Borowski, "This Way for the Gas, Ladies and Gentlemen," in *This Way for the Gas, Ladies and Gentlemen,* trans. Barbara Vedder (New York: Viking, 1967), p. 29.

19. Andrzej Wirth, "A Discovery of Tragedy: The Incomplete Account of Tadeusz Borowski," trans. Adam Czerniawki, *Polish Review* 12 (Summer 1967): 45.

20. Ibid., pp. 41, 35, 37. In some of my comments on Borowski, I draw on passages from my *Versions of Survival: The Holocaust and the Human Spirit* (Albany: State University of New York Press, 1982).

21. Sara Nomberg-Przytyk, *Auschwitz: True Tales from a Grotesque Land,* trans. Roslyn Hirsch (Chapel Hill: University of North Carolina Press, 1985), pp. 20–21.

22. Ibid., pp. 45–46.

23. Ibid., p. 111.

24. Ibid., p. 112.

25. Ibid., p. 113.

26. Ibid., pp. 153–154.

27. Primo Levi, *The Drowned and the Saved,* trans. Raymond Rosenthal (New York: Summit, 1988), p. 112.

28. Only *None of Us Will Return* exists in an English version. Rosette Lamont's translation of all three volumes will be published by Yale University Press in the spring of 1995. Her translation of *Days and Memory* (Marlboro, Vt.: Marlboro Press, 1990) is a separate volume, also partly concerned with Delbo's Auschwitz experience.

29. Charlotte Delbo, *None of Us Will Return,* trans. John Githens (New York: Grove Press, 1968), p. 20.

30. Ibid., p. 31.

31. Ibid.

32. Ibid., p. 122.

33. Charlotte Delbo, *Une Connaissance inutile* (Paris: Éditions de Minuit, 1970), pp. 183–184. [My translation]

34. Ibid., pp. 191, 190, 185.

Chapter 9

1. Franz Kafka, *Letter to His Father,* trans. Ernst Kaiser and Eithne Wilkins (New York: Schocken, 1966), p. 125. The text gives "Chinese puzzle" for *Geduld-spiel,* but the ingenious and intricate solution conveyed by this image seems less appropriate to Kafka's prolonged discussion of his relationship with his father than the literal allusion of his term to the card game of patience, akin to our solitaire.

2. George Steiner, *Language and Silence: Essays on Language, Literature, and the Inhuman* (New York: Atheneum, 1970), pp. 50, 121.

3. Ibid., p. 121.

4. "Challenges and Protests: Commentary by Bertolt Brecht," in *The World of Franz Kafka,* ed. J. P. Stern (New York: Holt, Rinehart and Winston, 1980), p. 180.

5. Günther Anders, *Franz Kafka,* trans. A. Steer and A. K. Thorlby (London: Bowes & Bowes, 1960), pp. 14–15.

6. Thorstein Veblen, "The Theory of Business Enterprise," in *The Portable Veblen,* ed. Max Lerner (New York: Viking, 1950), pp. 335, 336, 348.

7. Anders, *Franz Kafka,* p. 48.

8. Theodor W. Adorno, "Notes on Kafka," in *Prisms,* trans. Samuel and Shierry Weber (Cambridge, Mass.: MIT Press, 1981), pp. 245, 251–252.

9. Ibid., pp. 253, 255, 254.

10. Ibid., p. 259.

11. Ibid.

12. Ibid., pp. 259, 263.

13. Walter Benjamin, "Max Brod's Book on Kafka, and Some of My Own Reflections," in *Illuminations,* ed. Hannah Arendt, trans. Harry Zohn (New York: Harcourt, Brace & World, 1968), p. 146.

14. Ibid.

15. Anthony Thorlby, "Kafka and Language," in *World of Franz Kafka,* ed. Stern, p. 138.

16. *The Diaries of Franz Kafka: 1910–1913,* ed. Max Brod, trans. Joseph Kresh (New York: Schocken, 1965), pp. 291–292.

17. *The Diaries of Franz Kafka: 1914–1923,* ed. Max Brod, trans. Martin Greenberg with Hannah Arendt (New York: Schocken, 1965), p. 195.

18. Thorlby, "Kafka and Language," p. 141.

19. Jorge Semprun, *The Long Voyage,* trans. Richard Seaver (New York: Grove Press, 1964), p. 190.

20. *Diaries of Franz Kafka: 1914–1923,* p. 75.

21. Franz Kafka, *Letters to Friends, Family, and Editors,* trans. Richard Winston and Clara Winston (New York: Schocken, 1977), p. 208.

22. *Diaries of Franz Kafka: 1914–1923,* p. 92.

23. Kafka, *Letters to Friends*, p. 338.

24. Franz Kafka, *The Trial*, trans. Willa Muir and Edwin Muir, rev. E. M. Butler (New York: Schocken, 1968), p. 5.

25. Jakov Lind, "Soul of Wood," in *Soul of Wood and Other Stories*, trans. Ralph Manheim (New York: Fawcett Crest, 1966), p. 9.

26. Jean Améry, *Jenseits von Schuld und Sühne* (Munich: Szczesny, 1966), p. 33. [My translation]

27. Kafka, *Trial*, p. 228.

28. Martin Walser, "On Kafka's Novels," in *World of Franz Kafka*, ed. Stern, p. 87.

29. Franz Kafka, "The Metamorphosis," in *The Penal Colony: Stories and Short Pieces*, trans. Willa Muir and Edwin Muir (New York: Schocken, 1948), p. 124. Gregor's gradual dehumanization in the eyes of others is accompanied by a verbal demotion from "brother" to "creature" (*Untier*) to the charwoman's neuterized "thing" (*das Zeug*).

30. *Diaries of Franz Kafka: 1914–1923*, pp. 217–218.

Chapter 10

1. Aharon Appelfeld, *Badenheim 1939*, trans. Dalya Bilu (New York: Pocket Books, 1980), p. 53. [Subsequent references are in the text]

2. Peter Handke, "Repetition," *New Yorker*, 29 February 1988, pp. 34–35.

3. Aharon Appelfeld, *Tzili: The Story of a Life*, trans. Dalya Bilu (New York: Penguin, 1983), p. 3. [Subsequent references are in the text] In the original Hebrew, *Tzili* and *The Immortal Bartfuss* appeared in a single volume under the title *Hakootonet Vehapassim (The Gown and the Stripes)* (Tel Aviv: Hakibbutz Hameuchad, 1983).

4. Philip Roth, "A Talk with Aharon Appelfeld," *New York Times Book Review*, 28 February 1988, p. 31.

5. Aharon Appelfeld, *The Immortal Bartfuss*, trans. Jeffrey M. Green (New York: Weidenfeld & Nicolson, 1988), p. 61. [Subsequent references are in the text]

6. For a valuable study of Appelfeld's use of language in two other novels, *The Age of Wonders* and *Mikhvat Haor* (*Searing Light*), as well as important commentary on the literary challenges facing exiled writers, see Sidra Dekoven Ezrahi, "Aharon Appelfeld: The Search for a Language," *Studies in Contemporary Jewry* 1 (1984): 366–380.

Chapter 11

1. Cynthia Ozick, *The Shawl: A Story and Novella* (New York: Knopf, 1989), p. 14. [Subsequent references are in the text]

2. Fortunoff Video Archive for Holocaust Testimonies, Yale University, tape A-67: testimony of Bessie K.

Chapter 12

1. Albert Camus, *The Myth of Sisyphus and Other Essays,* trans. Justin O'Brien (New York: Vintage, 1955), p. 91.

2. Albert Camus, *The Plague,* trans. Stuart Gilbert (New York: Vintage, 1972), p. 287.

3. Quoted in Granville Hicks, "Literary Horizons," *Saturday Review,* 12 October 1963, p. 32.

4. Robert Alter, "Bernard Malamud: Jewishness as Metaphor," in Alter, *After the Tradition: Essays on Modern Jewish Writing* (New York: Dutton, 1969), pp. 125, 128.

5. Ibid., p. 118.

6. Bernard Malamud, *The Fixer* (New York: Farrar, Straus & Giroux, 1966), p. 333.

7. Ibid., p. 335.

8. Michael Brown, "Metaphor for Holocaust and Holocaust as Metaphor: *The Assistant* and *The Fixer* of Bernard Malamud Reexamined," *Judaism* 29 (Fall 1980): 484, 487.

9. Bernard Malamud, *The Assistant* (New York: Avon, 1980), p. 102.

10. Brown, "Metaphor," p. 484.

11. Ibid., p. 487.

12. Bernard Malamud, *The Magic Barrel* (New York: Avon, 1980), p. 171.

13. Ibid., p. 120.

14. Ibid., p. 144.

15. Ibid., p. 157.

16. Ibid., p. 162.

17. Brown, "Metaphor," p. 488.

Chapter 13

1. *Anne Frank: The Diary of a Young Girl,* trans. B. M. Mooyart-Doubleday (New York: Doubleday, 1967), p. 51.

2. Ibid., p. 287.

3. Ernst Schnabel, *Anne Frank: A Portrait in Courage,* trans. Richard Winston and Clara Winston (New York: Harcourt, Brace & World, 1958), p. 177.

4. Frances Goodrich and Albert Hackett, *The Diary of Anne Frank* (New York: Random House, 1956), p. 172.

5. Frank, *Diary,* p. 66.

6. Goodrich and Hackett, *Diary of Anne Frank,* p. 168.

7. Frank, *Diary,* p. 245.

8. Ibid.

9. Tadeusz Borowski, "This Way for the Gas, Ladies and Gentlemen," in *This Way for the Gas, Ladies and Gentlemen,* trans. Barbara Vedder (New York: Penguin, 1976), p. 45.

10. Millard Lampell, *The Wall* (New York: Knopf, 1961), pp. xiii, xiv, ix, xxiii.

11. Ibid., p. 158.
12. Ibid., pp. 99, 108, 124, 153.
13. Ibid., p. 156.
14. Arthur Miller, *Incident at Vichy* (New York: Viking, 1965), pp. 3, 6.
15. Ibid., pp. 19, 29, 30, 31, 33.
16. Ibid., pp. 38, 39, 46.
17. Ibid., pp. 50, 51.
18. Ibid., pp. 65, 66.
19. Abby Mann, *Judgment at Nuremberg: A Script of the Film* (London: Cassell, 1961), pp. 148, 180–181.
20. Ibid., pp. 149, 151.
21. Ibid., p. 153.
22. Elie Wiesel, "Trivializing the Holocaust: Semi-Fact and Semi-Fiction," *New York Times*, 16 April 1978, sec. 2, p. 29.
23. Lance Morrow, "Television and the Holocaust," *Time*, 1 May 1978, p. 53.
24. Jadwiga Bezwińska, ed., *Amidst a Nightmare of Crime: Manuscripts of Members of Sonderkommando*, trans. Krystyna Michalik (Oswiecim: State Museum of Oswiecim, 1973), p. 145.

Chapter 14

1. Aharon Appelfeld, "After the Holocaust," in *Writing and the Holocaust*, ed. Berel Lang (New York: Holmes & Meier, 1988), p. 92.
2. Terrence Des Pres, *The Survivor: An Anatomy of Life in the Death Camps* (New York: Oxford University Press, 1976), pp. 63, 192.
3. Frank Stern, *Jews in the Minds of Germans in the Postwar Period* (Paul Lecture, 1992) (Bloomington: Jewish Studies Program, Indiana University, 1993), p. 16.

Index